Amphibians, Reptiles, and
at Sabino Canyon

The Southwest Center Series Joseph C. Wilder, editor

Amphibians, Reptiles, and Their Habitats
at Sabino Canyon

David W. Lazaroff

Philip C. Rosen

Charles H. Lowe Jr.

The University of Arizona Press

Tucson

The University of Arizona Press
© 2006 The Arizona Board of Regents
All rights reserved
∞ This book is printed on acid-free, archival-quality paper.
Manufactured in the United States of America

Library of Congress Cataloging-in-Publication Data
Lazaroff, David Wentworth, 1948–
  Amphibians, reptiles, and their habitats at Sabino Canyon /
David W. Lazaroff, Philip C. Rosen, Charles H. Lowe, Jr.
    p. cm. – (The Southwest Center series)
  Includes bibliographical references and index.
  ISBN-13: 978-0-8165-2495-2 (pbk. : alk. paper)
  ISBN-10: 0-8165-2495-5 (pbk. : alk. paper)
    1. Amphibians—Arizona—Sabino Canyon. 2. Reptiles—
Arizona—Sabino Canyon. I. Rosen, Philip C. II. Lowe,
Charles H. III. Title. IV. Series.
QL653.A6L39 2006
597.909791′53—dc22
                              2005028021

FRIENDS OF **SABINO CANYON**

Publication of this book is made possible in part by grants from
the Friends of Sabino Canyon, the Sabino Canyon Volunteer
Naturalists, the Southwest Center of the University of Arizona,
and John W. Madden, M.D.

Maps for this book were prepared by TerraSystems Southwest,
Tucson.

Contents

Illustrations

Photographs

Color Plates   *following page 80*

Species Distribution Maps

Tables

Preface

Prolonged dry conditions . . . increase the likelihood and severity of wildfires
. . . if a wildfire were to burn a large area in the watershed of one of the
Recreation Area's two perennial streams, severe floods could follow . . .
with very heavy loads of ash and sediments.

When we wrote these words in an early draft of this book, we did not know
that just months later a century of fire suppression and several years of
drought would culminate in the largest and most destructive wildfire in
the Santa Catalina Mountains in living memory. Large parts of the watersheds of Sabino Creek and Bear Creek burned in June and July 2003,
and when the summer rains later fell on the denuded slopes, black water
rushed again and again down both streams, filling them with sediment
and debris. As we write this, it is still too early to know all the effects of
the Aspen Fire, but it is clear that some will be long lasting. For Sabino
Canyon, 2003 was a watershed year in more ways than one.

Our study of Sabino Canyon's amphibians, reptiles, and their habitats
began nearly a quarter century ago as an attempt to take an ecological
snapshot; it has since evolved into a moving picture of change. Flooding
was on the increase long before the events of summer 2003, populations of
native and introduced species have waxed and waned, suburbia has arrived
on the canyon's doorstep, and human visitation has soared. Change of another sort deeply affected the creation of this book, when our collaborator
and scientific mentor, Charles H. Lowe, passed away in 2002. Over his long
and influential career, Dr. Lowe had become increasingly concerned about
the careless degradation of the Sonoran Desert. Yet, despite his character-
istically vehement expression, "It's all over!" he thrived in his later years on
the continued struggle to sustain nature in the Southwest. Sabino Canyon
was one of many places he treasured. We dedicate this book to him.

David W. Lazaroff
Philip C. Rosen

Acknowledgments

The authors sincerely thank the many people who over more than twenty years shared sightings, offered technical assistance, and helped us in other ways far too numerous to describe: Craig Alexander, Diane Alexander, Robert Barnacastle, Michael Barry, Phil Bentley, Kyle Blasch, Fred Blatt, George Bradley, Danny Brower, Dennis Caldwell, Dennis Cornejo, Sarah Davis, Cindy Davison, Erik Enderson, Yvonne Endrizzi, Marion Erickson, Eldon Erwin, Elissa Fazio, Richard Felger, Barbara Franklin, Steve Hale, David Hall, Jonathan Hanson, June Hirsch, Peter Holm, Sue Hoyt, John Jacobs, Phil Jenkins, Terry Johnson, Cat Lazaroff, Bob Lefevre, John Wesley Miller, Gale Monson, Gerry Moore, Bill Mueller, Heather Murphy, Norma Niblett, Karen Nickey, Janice Perry, Steve Plevel, Peter Polley, Bob Porter, Brian Purnell, Roger Repp, Mindee Roth, Shawn Sartorius, Scott Schaller, Heidi Schewel, Cecil Schwalbe, Tom Skinner, Lin Smith, Margaret Strong, Dennis Suhre, Steve Taff, Josh Taiz, Joan Tedford, Rick Toomey, Peter Unmack, Eric Wallace, Darren Williams, Becky Wilson, Beth Woodin, Bill Woodin, and John Wright.

We gratefully acknowledge the following institutions for sharing records from their herpetological collections: The University of Arizona; American Museum of Natural History, New York; Academy of Natural Sciences, Philadelphia; Arizona State University; California Academy of Sciences; Carnegie Museum of Natural History, Pittsburgh; Field Museum of Natural History, Chicago; Illinois Natural History Survey; University of Kansas; Natural History Museum of Los Angeles County; Louisiana State University Museum of Natural Science; Museum of Vertebrate Zoology, University of California; San Diego Natural History Museum; University of Illinois Museum of Natural History; Museum of Zoology, University of Michigan; National Museum of Natural History, Smithsonian Institution; University of Texas at El Paso. Arizona State University also generously shared W. L. Mickley's fish database with us.

Our sincere thanks to the Coronado National Forest; the Department of Ecology and Evolutionary Biology, University of Arizona; and the Southwest Parks and Monuments Association for their support over the years. We owe a special debt of gratitude to the University of Arizona's

Southwest Center, the Friends of Sabino Canyon, the Sabino Canyon Volunteer Naturalists, and John W. Madden, M.D., for their generous financial support in bringing this work to publication. We would like to express our appreciation to Joe Wilder of the Southwest Center for his help and encouragement in reviving our long-dormant study.

Finally, special thanks to Cherie Lazaroff for her help in the field and for typing drafts more than two decades ago, in the primitive age before the word processor.

Amphibians, Reptiles, and Their Habitats at Sabino Canyon

Introduction

The Sabino Canyon Recreation Area lies in the cactus-studded Arizona Upland subdivision of the Sonoran Desert, at the foot of the Santa Catalina Mountains near Tucson, Arizona. Spanning the transition between the rugged mountain range and its surrounding skirt of sediments, the Recreation Area contains within its boundaries gently sloping plains, rocky foothills, and deep mountain canyons with permanent water. This varied topography fosters an impressive biological diversity, with arid desert communities in striking juxtaposition to lush riparian woodlands and semiarid grasslands.

Because of its beauty and its flowing streams, Sabino Canyon has been a favorite destination for Tucsonans for more than a century. Recreational facilities installed under federal relief programs during the Great Depression added greatly to its appeal, and in recent decades it has become increasingly popular among tourists. The Sabino Canyon Recreation Area received more than a million visitors in 2002, according to its administrators at the Santa Catalina Ranger District, Coronado National Forest (USDA Forest Service). The Recreation Area's roads have been closed to private automobiles since 1981, but it remains open to hikers, runners, bicyclists, and horseback riders, and a commercial shuttle operates daily from the entrance. The exclusion of private cars has helped greatly to protect plant and animal communities in this small natural area, but human pressures have inevitably increased with rising visitation and suburban encroachments on its boundaries.

There is a significant literature on visits to the Tucson area by early-twentieth-century herpetologists (scientists who study amphibians and reptiles), almost all of whom mention Sabino Canyon. Such notables as Alexander Ruthven (1907) and John Van Denburgh and Joseph Slevin (1913), who were among the first to do major herpetological work in Arizona, reported on their considerable time in the canyon. Other well-known figures, including Mary Dickerson, Laurence Klauber, K. P. Schmidt, and Charles Bogert, also collected in Sabino Canyon during the first half of the twentieth century, although unlike Willis King (1932) and Clinton McCoy (1932), they did not publish their observations. The museum records and published observations left by these careful investigators offer an invaluable view into the past.

Students and faculty at the University of Arizona have observed the amphibians and reptiles at Sabino Canyon for many years. Charles Lowe began studying the area in 1950. David Lazaroff and Lowe began intensive fieldwork for this report in early 1980 and continued it during all seasons into 1983. In addition to hiking throughout the area, we employed pitfall traps, repeat spot sampling, and road hunting from vehicles. We searched the University of Arizona Herpetology Collection for specimens from the Recreation Area, preserved and deposited some additional voucher material, and photographed and released selected animals. We also inventoried and photographed the Recreation Area's major biotic communities and environments.

After 1983 we made occasional observations in the Recreation Area until 2001, by which time Lowe was unfortunately no longer able to participate, due to declining health. Philip Rosen, who had first visited the Recreation Area in 1986, then joined the project. He and Lazaroff returned to the Recreation Area repeatedly during 2002 and 2003 to reassess the status of the herpetofauna, with special attention to the amphibians. To better understand the wider context of certain populations, we explored Sabino Creek and Bear Creek north and south of the Recreation Area boundaries. We searched for additional specimens and records in computer databases of the University of Arizona Herpetology Collection and other museum collections in the United States, and we solicited recollections of other observers with experience in and near the Recreation Area, particularly prior to the 1980s. We revisited all the sites of earlier environmental photographs and repeated many to assess ecological change. (Space permits publishing only a few of the repeat photographs here, but all photographs will be archived for future investigators.)

Our objective has been to present an annotated and illustrated checklist of the herpetofauna within the environmental framework of the topographic features and biotic communities of the Sabino Canyon Recreation Area. We hope the information in these pages will be useful to those who seek to understand the natural history of this beautiful desert area, how it has changed, and how it may change in the future.

Safety for Herptiles and Humans

Sabino Canyon's herptiles (amphibians and reptiles) are eye-catching creatures. Like birds, they are sometimes colorful and often busily moving about. However, unlike birds they can be temptingly easy to catch—a difference that can lead to problems both for the animals being caught and for the people doing the catching. We encourage visitors to Sabino Canyon and other natural areas to enjoy wild amphibians and reptiles much as they do wild birds. A hands-off approach of "herp watching," with the aid of binoculars, opens the door to a wide range of fascinating animal behavior, but that door slams shut once an animal is caught. Moreover, we need to recognize that our presence in natural areas like Sabino Canyon has become so frequent that even such apparently innocuous interference can be harmful to the wildlife we enjoy.

For those who feel they must catch an amphibian or reptile, certain facts should be kept in mind. All Sabino Canyon's frogs and toads protect themselves to some degree by means of toxic skin secretions. Secretions from a canyon treefrog or a spadefoot toad may burn for hours if accidentally rubbed in your eyes. Sonoran Desert (Colorado River) toads regularly poison Tucson-area dogs, as is well known, and their secretions are potentially dangerous to humans as well. Small children should not handle these animals. If you touch any frog and toad, wash your hands afterward.

Although many of Sabino Canyon's reptiles bite vigorously if caught, only a few are dangerously venomous: the Gila monster, the Sonoran coralsnake, and the area's several species of rattlesnakes. Common sense dictates that none of these should be picked up or closely approached. Nor should they be harmed; each is an interesting animal with a role to play in its world. Reasonable care makes it unlikely that a visitor will be injured by any reptile that he or she has not deliberately molested. Nevertheless, the danger of a rattlesnake bite, while minimal, is real. Watch where you put your hands and feet, use a flashlight after dark, and familiarize yourself with first-aid procedures (for example, see Lowe et al. 1986; Hare 1995; Lazaroff 1998).

If you do catch an amphibian or reptile, gentle and brief handling will minimize its stress (yet significant stress is inevitable and may be repeated later at the mere sight of another visitor). After examining the animal, put it back exactly where you caught it. It has a well-defined home range from which it rarely strays, and it may die if released elsewhere. Never remove an amphibian or reptile (or any other animal or plant) from the Sabino Canyon Recreation Area, and if you see someone else doing so, report this to Forest Service personnel. Poaching of amphibians and reptiles is a serious problem in many desert areas, and your vigilance can help keep it from becoming one at Sabino Canyon. A special regulation published by Coronado National Forest protects all the Recreation Area's animals and plants from harm or removal, and many species are additionally protected by the State of Arizona.

Finally, be aware that releasing unwanted pet amphibians and reptiles in Sabino Canyon is dangerous both to the pets themselves and to the area's native wildlife. This all-too-common practice is a serious problem in the Recreation Area. Exotic turtles found with chewed-off legs in Lower Sabino Canyon are the lucky ones—most released pets do not survive. Worse yet, released pets might thrive and reproduce in the Recreation Area or bring with them exotic diseases, either of which could have disastrous consequences for native species. For these reasons, since 1984 it has been against Forest Service regulations to release any animal in the Sabino Canyon Recreation Area; such releases are also a violation of state law. The worldwide problem of introduced animals and diseases is severe, though not yet fully recognized by citizens and governments.

Part I The Ecological Setting

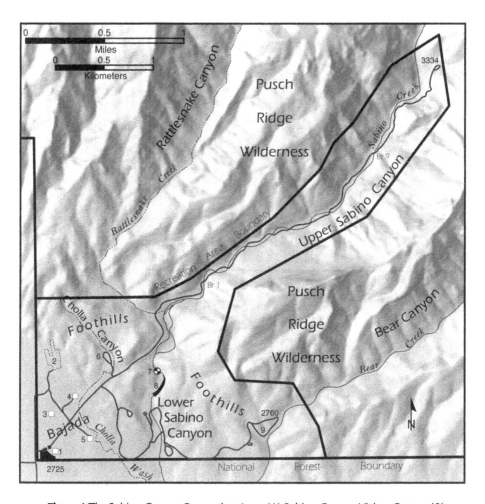

Figure 1 The Sabino Canyon Recreation Area: (1) Sabino Canyon Visitor Center; (2) shooting range; (3) warehouse; (4) shuttle service area; (5) Lowell Administrative Site; (6) Cactus Picnic Area; (7) old gauging station; (8) Sabino Lake; (9) Lower Bear Picnic Area (abandoned).

Topographic Features and Localities

The Santa Catalina Mountains, which reach an elevation of 9157 ft. at Mount Lemmon, are one of the biologically diverse sky-island ranges that rise above the deserts and grasslands of southeastern Arizona. The Sabino Canyon Recreation Area (fig. 1), named for the deep canyon that is its most important feature, is located at the base of the range's southern slope. Elevations in the Recreation Area range from 2660 to 3800 ft. The 2.2-square-mile (1400-acre) area divides naturally into six topographic regions: bajada, foothills, Upper Sabino Canyon, Lower Sabino Canyon, Rattlesnake Canyon, and Bear Canyon.

The scientific names of plants and animals (other than amphibians and reptiles) mentioned in this section and throughout this book are given in Appendix A.

Bajada

A small section of the gently sloping southern bajada (alluvial outwash plain) of the Santa Catalina Mountains is included in the southwestern corner of the Recreation Area (fig. 2), including a finger extending northeastward into Cactus Picnic Area. Numerous small drainages, the largest of which is Cholla Wash (see fig. 13), cross the bajada and eventually join Sabino Creek south of the Recreation Area boundary.

Foothills

In the Recreation Area a region of low, rocky foothills stands between the bajada and the steep front of the Santa Catalina Mountains. This topographic unit includes the hills north of the shuttle service area and southwest of Rattlesnake Canyon, the low ridge west of Lower Sabino Canyon, and the hilly terrain between Lower Sabino Canyon and Bear Canyon. Some of the minor washes that cross the bajada originate in the foothills; an example is Cholla Wash, which begins in Cholla Canyon (figs. 3a, b) northwest of Cactus Picnic Area.

Upper and Lower Sabino Canyon

Sabino Canyon, which begins near the summit of the Santa Catalinas, is the largest drainage on the southern slope of the mountain range. Although the uppermost reaches of the canyon lie far above the Recreation

Area, and Sabino Creek extends across the bajada several miles below the Recreation Area, we conform to local custom in using the terms "Upper Sabino Canyon" and "Lower Sabino Canyon" to refer to parts of the canyon lying entirely within the Recreation Area boundaries.

Upper Sabino Canyon (figs. 4, 5) includes most of the roughly 3-mi. portion of Sabino Canyon within the Recreation Area. Its steep and spectacular walls, incised by numerous side drainages, rise in some places more than 1500 ft. above the canyon floor. In contrast, Lower Sabino Canyon (fig. 6), which is essentially the canyon mouth, is bordered by the low foothills described above. For convenience, we have adopted the site of the old water gauging station above Sabino Lake as the boundary between Upper and Lower Sabino Canyon. (This long-dismantled station was replaced in 1981 with one in the lake itself.)

The paved road in Upper Sabino Canyon (fig. 7) is the most traveled route in the Recreation Area and so provides the setting in which the greatest number of visitors encounter amphibians and reptiles.

Rattlesnake Canyon

Rattlesnake Canyon (fig. 8) is an important tributary of Upper Sabino Canyon, which it joins 0.4 mi. upstream from the old gauging station. Only the lowest 1200 ft. of this mountain canyon are within the Recreation Area boundaries.

Bear Canyon

Like Sabino Canyon, Bear Canyon (fig. 9) is a steep-sided mountain canyon and a major topographic feature of the southern slope of the Santa Catalina Mountains. Less than 0.5 mi. of Bear Canyon is in the Recreation Area, including the broad bench formerly occupied by Lower Bear Picnic Area. Bear Creek joins Sabino Creek 0.9 mi. south of the Recreation Area boundary. The popular Bear Canyon Trail (fig. 10), often called the "Seven Falls Trail," leads up the canyon beyond the Recreation Area boundary.

Figure 2 Bajada of the Santa Catalina Mountains, October 1980. Looking southwestward toward Tucson from foothills near Cactus Picnic Area, camera at 2840 ft. Bajada desertscrub, here primarily a velvet mesquite woodland, covers gently sloping bajada in middle distance. Foreground plants, including barrel cactus between two ocotillos, are in paloverde-saguaro desertscrub of foothills. Distant houses and hills lie beyond the Recreation Area boundary. Since this photo was taken, additional commercial and residential developments have contributed to increased recreational visits, intrusions by neighborhood pets, potential for exotic plant invasions, and isolation of certain native animal populations, such as the bajada-dwelling leopard lizard.

Figure 3a Foothills of the Santa Catalina Mountains, October 1980. Looking northwestward up Cholla Canyon from near Cactus Picnic Area toward southern escarpment of range, camera at 2860 ft. Washes of this sort are avenues along which species characteristic of the bajada, such as zebra-tailed lizard and tiger whiptail, penetrate the foothills environment. Upland community is paloverde-saguaro desertscrub, but catclaw, white thorn, velvet mesquite, and coursetia are also important trees on the middle-distance slopes and (except for coursetia) on terraces bordering the wash.

Figure 3b Foothills of the Santa Catalina Mountains, October 2002. Saguaro population has declined noticeably in density, and most conspicuous change has been loss of many of the largest plants. Trees and shrubs have grown, increasing perennial vegetative cover on the slopes, and Engelmann prickly pear numbers have greatly increased throughout. Similar changes in the paloverde-saguaro community have been documented elsewhere in the Recreation Area and may reflect regional trends (Turner et al. 2003: 186–91, 248–55). In contrast to slopes, many trees on terrace to right of wash have died, decreasing cover, perhaps in response to drought in recent years.

Figure 4 Upper Sabino Canyon, October 1982. Looking northeastward up canyon from slope above first bridge, camera at 3100 ft. Thimble Peak, elevation 5323 ft., on skyline. Paloverde-saguaro desertscrub on northwestern canyon wall (left), ecotone between paloverde-saguaro desertscrub and semidesert grassland on southeastern wall (right). Riparian woodland and mesquite bosque compose dense vegetation on the canyon floor. Ranges of several reptiles characteristic of higher elevations, such as Great Plains skink and ring-necked snake, finger downward into Recreation Area in these riparian communities. The 1995 Sabino Fire ignited near bottom center and spread upslope to right, then up canyon until stopped short of Thimble Peak. Aspen Fire, centered higher in the mountain range, burned the ridgetop near Thimble Peak in 2003.

Figure 5 Upper Sabino Canyon, October 1982. Looking northward up canyon from near old gauging station, camera at 3000 ft. Rattlesnake Canyon joins Sabino Canyon at upper left. Sabino Creek divides to form an elongated island here, with primary channel at foot of slope below the road, secondary channel below the camera. The broad canyon floor supports especially extensive and diverse riparian scrub and mesquite bosque communities. Mexican blue oak, Emory oak, and other trees characteristic of higher elevations are unusually common in the riparian woodland here (though oak populations, already declining in 1982, have continued to do so). Regal horned lizard distribution reaches up the canyon into this area; canyon spotted whiptail is often seen here.

Figure 6 Lower Sabino Canyon, October 1982. Looking southwestward down canyon from near old gauging station, camera at 3000 ft. Animals characteristic of both bajada and canyon environments are found together here in the most diverse local herpetofauna in the Recreation Area. In foreground, riparian forest on silted-in bed of Sabino Lake, bordered on right (west) by narrow mesquite bosque. Tall trees of riparian woodland mark channel of Sabino Creek extending beyond lake toward upper left. Paloverde-saguaro desertscrub on hills; bajada desertscrub in distance, south of Recreation Area boundary, was largely disturbed or replaced by residential developments by 2002. Some roads south of lake were abandoned or moved following severe flood damage in 1993.

Figure 7 Sabino Canyon Road, Upper Sabino Canyon, November 1980. Looking northeastward up canyon from between eighth and ninth bridges, camera at 3020 ft. Shrubby roadside vegetation consists of canyon ragweed, coursetia, desert broom, and hop bush. Species most often seen on the pavement by day are greater earless lizard, Sonoran spotted whiptail, and canyon spotted whiptail; Sonoran whipsnake and black-tailed rattlesnake are less common. At night these are replaced by western lyresnake, tiger rattlesnake, and, especially during and after summer rainstorms, red-spotted toad and Sonoran Desert toad.

Figure 8 Rattlesnake Canyon, November 1980. Looking northwestward up Rattlesnake Creek toward the boundary of the Recreation Area, camera at 2800 ft. The intermittent canyon stream is dry, as is usually the case. Among the principal plants in riparian scrub along the channel are velvet mesquite, canyon ragweed, catclaw, coursetia, and Engelmann prickly pear. Paloverde-saguaro desertscrub grows on the slopes. Greater earless lizard, Sonoran spotted whiptail, canyon spotted whiptail, tree lizard, and Clark's spiny lizard are frequently seen here. Canyon treefrog and black-necked garter snake appear when the stream flows. Several flash floods occurred here after the 2003 Aspen Fire, but mud was scoured out by the last flood, leaving the environment little changed.

Figure 9 Bear Canyon, October 1980. Looking northeastward up canyon past Lower Bear Picnic Area in foreground, camera at 2840 ft. Paloverde-saguaro desertscrub on slopes in foreground and middle distance. Dense mesquite bosque in foreground center, bordered on right by riparian woodland along channel at foot of slope. Amphibians and reptiles characteristic of canyon and bajada environments meet in former Lower Bear Picnic Area as in Lower Sabino Canyon, though the fauna here is less diverse. By 2002 the picnic area had been abandoned and the road in foreground revegetated.

Figure 10 Bear Canyon Trail, October 1980. Looking northeastward up Bear Canyon toward the Pusch Ridge Wilderness from near Lower Bear Picnic Area, camera at 2760 ft. Paloverde-saguaro desertscrub on slope at left. Sparse riparian woodland along dry channel of Bear Creek at center. Mesquite bosque on terrace at right. Hikers frequently encounter Sonoran spotted whiptail, canyon spotted whiptail, and greater earless lizard on trail; tree lizard and Clark's spiny lizard on rocks nearby. Canyon treefrog and black-necked garter snake are seen along channel where surface water is present. Tiny, freshly transformed red-spotted toads may be briefly abundant in late summer. Floods following the Aspen Fire deposited finer sediments along the stream in 2003.

Climate

Sabino Canyon's climate reflects the Southwestern pattern of biseasonal rainfall. Precipitation arrives mostly in summer and winter, typically with intervening periods of spring and fall drought. Summer rainfall (May–October) accounts for 56% of the total and is mostly derived from monsoon thunderstorms in July, August, and September. The remaining 44% (winter precipitation, November–April) falls primarily during gentler storms from December through March. Snowfall is infrequent in the Recreation Area and occurs mostly at higher elevations on the walls of Sabino Canyon and Bear Canyon. The coolest month is January; the warmest, July.

Table 1 lists statistics for the Sabino Canyon weather station, located just south of the National Forest boundary, 0.5 mi. east of the Sabino Canyon Visitor Center. Figures given are climatological normals for 1951–1980. Data from this station are not available after 1982.

Table 1 Climate at the Sabino Canyon weather station, elevation ca. 2640 ft.

	Jan	Feb	Mar	Apr	May	Jun	Jul	Aug	Sep	Oct	Nov	Dec	Year
Mean minimum temperature (°F)	37.1	39.0	42.8	48.6	56.3	65.9	72.4	70.2	65.7	54.9	43.7	37.8	52.9
Mean maximum temperature (°F)	66.8	69.8	74.2	82.6	91.5	101.4	101.9	99.5	96.8	87.1	75.0	67.9	84.5
Total precipitation (in.)	1.26	0.91	1.09	0.46	0.17	0.24	2.35	2.31	1.17	1.06	0.82	1.26	13.10

Source: National Oceanic and Atmospheric Administration (1982)

Terrestrial Environments

To describe the ecological distributions of the amphibians and reptiles, we recognize six terrestrial biotic communities within the Sabino Canyon Recreation Area: bajada desertscrub, paloverde-saguaro desertscrub, semidesert grassland, riparian woodland and forest, riparian scrub, and mesquite bosque. Numbers following the community names refer to the digitized classification system of Brown et al. (1980). The Recreation Area's terrestrial communities have been mapped by Lazaroff (1993: 6-7).

Bajada Desertscrub (154.1)
The Sonoran desertscrub community on the bajada in the Recreation Area is a mosaic controlled by soil characteristics and minor variations in topography. Much of the vegetation is an open mesquite woodland with a variable understory dominated by one or more small shrubs (figs. 11a, b). There are also small areas dominated by creosotebush (fig. 12).

The desert wash (fig. 13), an environment within bajada desertscrub, supports higher densities of certain plants, such as velvet mesquite and catclaw, due to enhanced soil moisture. This vegetation is often called "xero-riparian" (dry riparian).

The large buildings in the Recreation Area—the Sabino Canyon Visitor Center and office complex, Lowell Administrative Site, and the Coronado National Forest warehouse—are all on the bajada. The Tucson Rod and Gun Club's shooting range, currently in disuse, includes smaller structures and large areas of disturbed and barren ground on the bajada.

Paloverde-Saguaro Desertscrub (154.1)
This characteristic Arizona Upland community is found in the Recreation Area on the shallow, rocky soils of the foothills (figs. 3a, b) and the walls of Sabino, Bear, and Rattlesnake Canyons. In Upper Sabino Canyon paloverde-saguaro desertscrub is most extensively developed on the northwestern (southeast-facing) canyon wall (fig. 14). Whereas the dominant plants in this community—the foothill paloverde and the saguaro—are the same in the foothills and in the canyons, the two topographic regions differ substantially in herpetofauna.

Figure 11a Bajada desertscrub, dominated locally by burrobrush, October 1980. Looking northward toward the southern slope of the Santa Catalina Mountains from near Lowell Administrative Site, camera at 2720 ft. Jumping cholla in foreground right; barrel cactus with crown of fruits in foreground center. The many small shrubs across the foreground are almost all burrobrush, and the darker trees in middle distance are velvet mesquites. Similar communities on the bajada are dominated locally by burroweed or triangle bur-sage. Amphibians and reptiles as in fig. 12.

Figure 11b Bajada desertscrub, dominated locally by triangle bur-sage, October 2002. Shrubby foreground vegetation may seem little changed, but in fact nearly complete species replacement has occurred. Burrobrush plants have mostly died or been reduced to a few living stems, and triangle bur-sage plants, many of which took root beneath the burrobrush canopies, now dominate. (Shift may be related to drought; burrobrush still thrives along small drainages nearby.) Jumping cholla and barrel cactus in earlier photograph have died, but others have appeared, and Engelmann prickly pear has greatly increased. Saguaros have overtopped velvet mesquite nurse plants, which have also grown in height. However, mesquites have many dead branches and are heavily parasitized by desert mistletoe, and some trees have died.

Figure 12 Bajada desertscrub, dominated locally by creosotebush, October 1980. Looking northward from near Visitor Center toward Santa Catalina Mountains, camera at 2720 ft. Small white thorn in foreground right. Saguaro is associated with velvet mesquite nurse tree. (By 2002 only a stump and a standing skeleton remained of the white thorn and the saguaro.) Tiger whiptail, side-blotched lizard, and zebra-tailed lizard are frequently seen here by day. Red-spotted toad and Couch's spadefoot are active on wet nights. The most commonly seen snakes are gopher snake, coachwhip, and western diamondback rattlesnake.

Figure 13 Cholla Wash, October 1980. Looking northeastward (upstream) near the shuttle service area, camera at 2740 ft. Velvet mesquite at left above Engelmann prickly pear. Canyon ragweed at right, behind it a large desert hackberry. Dry sandy beds of this and other washes on the bajada are primary natural habitat for zebra-tailed lizard. Tree lizard and desert spiny lizard live among mesquites along banks; other animals listed in figure 12 are found here also. Flow in this ephemeral desert stream is too brief to support amphibian reproduction.

Figure 14 Paloverde-saguaro desertscrub on the northwestern wall of Upper Sabino Canyon, January 1983. Looking westward from Sabino Canyon Road between eighth and ninth bridges, camera at 3040 ft. Teddy bear cholla in foreground center. Foothill paloverde, foreground right, is principal large shrub in this community; brittlebush and coursetia are also important. Sonoran spotted whiptail and canyon spotted whiptail can be found foraging on the slope, with tree lizard and Clark's spiny lizard basking on rock outcrops, though all are less abundant here than on the opposing canyon wall (fig. 15).

Figure 15 Ecotone between paloverde-saguaro desertscrub and semidesert grassland on the southeastern wall of Upper Sabino Canyon, January 1983. Same camera station as for figure 14, but turned roughly 180°, looking eastward up side
drainage on opposite canyon wall. Saguaros are mostly confined to warmer south-
and southwest-facing slopes on left side of drainage and diminish in density with
increasing elevation. Both blue and foothill paloverde are found here, the former
more abundant. Arizona rosewood dots upper slopes near cliffs. Ridgetop is 5000
ft. Secretive Great Plains skink and western banded gecko inhabit the drainage;
other reptiles as listed in figure 14. Area of photo lies near northern limit of the
June 1995 Sabino Fire.

Semidesert Grassland (143.1)

Within the Recreation Area semidesert grassland is found only on the southeastern wall of Upper Sabino Canyon, which, because of its more northerly-facing aspect, is a cooler, less-arid environment than the opposing slope. Figures 14 and 15, taken from the same camera station, illustrate the contrasting vegetation of the two canyon walls.

The vegetation on the southeastern canyon wall is not entirely semidesert grassland, but rather a complex desertscrub/grassland ecotone (ecological transition zone) in which desertscrub plants predominate on the more southerly-facing slopes and semidesert grassland plants predominate on the more northerly-facing slopes of the numerous ridges and side canyons. Plant species diversities on such less-arid, lower-elevation slopes in the Santa Catalina Mountains have been described as among the highest in the United States (Whittaker and Niering 1965: 449). The Phoneline Trail (fig. 16) provides a convenient transect through this interesting environment.

Riparian Woodland and Forest (223.2)

A distinctive community dominated by tall, broad-leaved, winter-deciduous trees grows along the channels of Sabino Creek and Bear Creek. Within the Recreation Area this community is mostly an open-canopied woodland (figs. 17a, b); only here and there along Sabino Creek does the vegetation achieve the closed canopy that justifies the designation "forest." A unique example of true riparian forest exists today on the sediment that now mostly fills the bed of Sabino Lake (figs. 18a, b) in Lower Sabino Canyon.

Riparian Scrub (233.2)

The discontinuous sandy floodplains found next to and at levels slightly above the channels of Sabino Creek and Bear Creek typically support a biotic community dominated by shrubs and small trees. In the Recreation Area this riparian scrub community (fig. 19) is best developed in the broader reaches of Sabino Canyon between Rattlesnake Creek and the southern boundary. A similar community is found along the channel of Rattlesnake Creek (fig. 8).

Mesquite Bosque (224.5)

This frequently dense mesquite-dominated woodland (fig. 20) is found chiefly in Sabino Canyon and Bear Canyon, on terraces above the levels of the sandy floodplains. Much of the mesquite bosque that formerly existed

Figure 16 Semidesert grassland on the southeastern wall of Upper Sabino Canyon, October 1980. Looking southwestward toward the Phoneline Trail, camera at 3600 ft. A community largely of shrubs and perennial grasses, on a north-facing slope. Arizona rosewoods conspicuous at left, with ocotillos and sotols nearby. Among many other woody perennials here are Wright lippia, fairy duster, turpentine bush, and catclaw. (By 2002 hop bush, uncommon here in 1980, had proliferated across this unburned slope, north of the June 1995 Sabino Fire.) Tree lizard and Clark's spiny lizard usually seen on rocks near trail; Sonoran spotted whiptail, canyon spotted whiptail, and greater earless lizard on the trail itself. Sonoran whipsnake and tiger rattlesnake occasionally encountered.

Figure 17a Riparian woodland and perennial canyon stream, May 1981. Looking southward (downstream) along channel of Sabino Creek from highest bridge, Upper Sabino Canyon, camera at 3080 ft. Large trees here are Bonpland willow (left), velvet ash (angular trunk behind boulder at center), Arizona sycamore (right), Fremont cottonwood, and Arizona walnut. Clark's spiny lizard, tree lizard, and Sonoran whipsnake are found here among boulders and in trees; canyon treefrog on smooth rocks near stream; black-necked garter snake in or near water. Sonoran mud turtle is still seen infrequently, but lowland leopard frog, abundant in the early twentieth century, was absent by this date. Exotic green sunfish invaded this far upstream by 1995, largely replacing native Gila chub.

Figure 17b Riparian woodland and perennial canyon stream, May 2002. Effects of changes in stream flow patterns are readily apparent. Floodwaters have scoured channel and moved large boulders formerly at right. Bonpland willow formerly at left is gone, but saplings have grown across foreground. Rocks have partly protected some trees; velvet ash behind central boulder has broken off and is vigorously resprouting, out of view of camera. Other trees, less protected, have resprouted from badly eroded stumps. Flood debris is piled against base of the Arizona sycamore to 6 ft. height, and bark is scarred at 10 ft.; however, loss of upper branches may be due to drought. Exotic fountain grass, present outside frame in 1981, has spread onto shore, right of center.

Figure 18a Riparian forest on sediment-filled bed of Sabino Lake, Lower Sabino Canyon, October 1980. Looking eastward, camera at 2720 ft. Community dominated by large Goodding willows, grown from wind-borne seeds that sprouted as pioneer seedlings on the sediments. Smaller velvet ash and Arizona walnut trees are less important constituents, as is net leaf hackberry. A few small African sumacs and a clump of giant reed (both species exotic) are growing at southwestern edge of lakebed, outside photograph. Thickets of canyon ragweed, common cocklebur, and sacred datura provide escape cover for ground-dwelling canyon spotted whiptail and Sonoran spotted whiptail. Clark's spiny lizard and tree lizard occupy the trees.

Figure 18b Riparian forest on sediment-filled bed of Sabino Lake, October 2002. Two decades of ecological succession and sporadic severe flooding have so altered this environment that earlier camera station can be relocated only within a few yards. The forest is both denser and more diverse, and its floor is hummocky and choked with debris. Many senescent Goodding willows have fallen; large branches at upper right have grown from a fallen trunk. Arizona walnut and velvet ash trees are now mature and reproducing. Vertical ash stem left of center arises from a sapling bent horizontally by floodwaters. Dark foliage right of center is African sumac, which has spread across lakebed, as has giant reed (not shown). Other plants listed for 1980 (and many others) still present.

Figure 19 Riparian scrub on the floodplain of Sabino Creek, October 1980. Lower Sabino Canyon, looking southward (downstream) from below the lower bridge, camera at 2680 ft. Tall trees in background are velvet ash and Goodding willow in riparian woodland along channel. Foreground plants include deer grass at far left, wait-a-minute bush at center, desert broom at right. Nearly leafless branch at right is velvet mesquite at edge of mesquite bosque, weakened by flooding. (By 2002 continued flooding had eroded several yards from the bosque and reduced the tree to a stump.) Greater earless lizard, side-blotched lizard, tree lizard, and three species of whiptail lizards are commonly seen here by day, red-spotted and Sonoran Desert toads at night.

Figure 20 Mesquite bosque on a stream terrace in Lower Bear Picnic Area, Bear Canyon, November 1980. Looking southeastward, camera at 2760 ft. Dead stems of exotic annual London rocket in front of conspicuous velvet mesquite, foreground left. Small gum bumelia at far right, behind it a thicket of desert hackberry. Among other shrubs and trees important in this community are catclaw, canyon ragweed, and blue paloverde. Tree lizard, Clark's spiny lizard, and the three species of whiptails are often seen here; regal horned lizard, gopher snake, and Sonoran whipsnake less often. Red-spotted toad metamorphs may be briefly abundant (as in fig. 10).

in the canyons has been greatly disturbed or destroyed by the construction of roads and picnic areas. The best remaining examples are found today on the wide bench formerly occupied by Lower Bear Picnic Area (fig. 9) and on the broad floor of Upper Sabino Canyon between Rattlesnake Canyon and the old gauging station (fig. 5). There is also a small mesquite bosque outside the two major canyons, along a short stretch of Cholla Wash just below Cactus Picnic Area.

Aquatic Environments

For convenience in describing the ecological distributions and breeding habitats of the amphibians and reptiles, we have divided the Recreation Area's important aquatic environments into four categories: perennial canyon stream, intermittent canyon stream, ephemeral desert stream, and ephemeral desert pool.

Perennial Canyon Stream

Sabino Creek and Bear Creek, on the floors of the Recreation Area's two largest mountain canyons, both have perennial underflow, though neither has perennial surface flow along its entire length. In Bear Creek there is no perennial surface water at all in the short reach within the Recreation Area boundaries, though there are perennial pools farther upstream (along the Bear Canyon Trail, fig. 10). In Sabino Creek there are several short perennially flowing reaches within the Recreation Area itself, and until recently there were also numerous rocky- and sandy-bottomed perennial pools (fig. 21a). Such dependable waters have significance for the native amphibians both as breeding sites and because they have allowed the persistence of certain aquatic predators, in particular the native Gila chub and the exotic green sunfish, northern crayfish, and American bullfrog. However, flash flooding following the large Aspen Fire in 2003 reduced the number of perennial pools (fig. 21b) and directly affected these predators. (We discuss these phenomena in more detail in the Ecological Change chapter.)

Sabino Creek's headwaters are in fir forest at an elevation of approximately 9000 ft., and the stream drains a 36-square-mile watershed. Within the Recreation Area many reaches have surface flow for months at a time, though much of the channel often becomes dry during the drought periods directly before and after the summer rains. Bear Creek's headwaters are in pine forest at approximately 8000 ft. It drains a smaller watershed than Sabino Creek and consequently flows visibly for shorter periods. Both streams are subject to occasional flooding during both the summer and winter rainy seasons. The streams then spread beyond their boulder-strewn channels across the sandy lower floodplains and occasionally onto the higher stream terraces, where they may damage roads and recreational facilities.

Figure 21a Perennial pool in Sabino Creek, October 1980. Lower Sabino Canyon, looking northward toward the dam at Sabino Lake (background right), camera at 2700 ft. At date of photo four exotic aquatic animals inhabited this large pool: northern crayfish, mosquitofish, green sunfish, and American bullfrog, though the frog, common here in the 1960s, had already been greatly reduced in numbers by violent flooding. Until 2003 native Sonoran Desert toads regularly bred here during the summer monsoon, though there was little successful reproduction.

Figure 21b Formerly perennial pool in Sabino Creek, June 2004. Intense late-twentieth-century floods have shifted boulders, but most significant change is burial of pool by flood-borne sediments after the 2003 Aspen Fire. On this day in the arid foresummer only shallow puddles temporarily remain at perimeter of the sand. Mosquitofish was wiped out by a flood in 1993; flooding after the Aspen Fire extirpated green sunfish and decimated northern crayfish, but American bullfrogs continue to immigrate from downstream. By 2004 most of the larger pools in Sabino Creek were no longer perennial, but smaller perennial pools remained, mostly in the upper canyon, above the eighth bridge.

Man-made structures dating from the 1930s have significantly altered Sabino Creek. In Upper Sabino Canyon the road crosses the stream nine times over small stonework dams (today usually called "bridges"). Even before the Aspen Fire, sandy sediments had collected behind some of these structures for several hundred feet. Shallow ponds cover these deposits during moderate streamflow, but many of these sites are among the places where water first disappears beneath the sand during times of drought. Two smaller bridges across the main stream channel in Lower Sabino Canyon have collected lesser quantities of sediment.

Sabino Lake (fig. 6), created by construction of a larger recreational dam in Lower Sabino Canyon during the 1930s, has mostly filled with sandy sediments. Now greatly reduced in surface area, it exists today at times of moderate streamflow as an elongated, shallow pond, bordered on both sides by a dense riparian forest growing on sediment collected by the dam. During low streamflow the pond's surface mostly recedes beneath the sand. Until recently Sabino Lake was perennial, but only in the strictest sense; it was sometimes reduced to a puddle a few yards across. Sediments deposited after the Aspen Fire changed this, and for a time during the dry foresummer of 2004 there was no surface water anywhere in the lake. Remnants of several much smaller recreational dams and their accumulated sediments are scattered throughout Upper Sabino Canyon.

Small bedrock potholes found here and there along Sabino Creek, perched above the usual stream level, are often filled during episodes of high water and by local rainwater runoff. These basins are usually free of fish and crayfish. Shallow temporary ponds in secondary (overflow) channels of Sabino Creek (figs. 22a, b), mostly in the broader areas of the canyon floor below the Rattlesnake Creek confluence, are sometimes filled by episodic high flow, local runoff, or seepage from the primary channel. These, too, are sometimes free of fish and crayfish.

Intermittent Canyon Stream

Rattlesnake Creek (fig. 8), on the floor of the Recreation Area's third mountain canyon, has its headwaters at an elevation of approximately 6000 ft. and drains a relatively small watershed. The boulder-lined channel is usually dry, and the stream flows persistently only after periods of exceptional rainfall. Bedrock plunge pools (tinajas) near the Recreation Area boundary fill at these times, and at other times they may contain water from local runoff; however, there is no perennial water in Rattlesnake Creek in or near the Recreation Area. Downstream from the plunge pools, the intermittent stream has occasionally been invaded by crayfish and fish from Sabino Creek.

Ephemeral Desert Stream

Water flows in the sandy desert washes in the foothills (figs. 3a, b) and on the bajada (fig. 13) only briefly during and following heavy local rainfall. Such ephemeral flow, typically lasting less than an hour, most often results from summer thunderstorms. It has little significance for amphibian reproduction in the Recreation Area. However, tadpoles have survived to metamorphosis in persistent pools in Cholla Wash, near the southern Recreation Area boundary, where water from ephemeral flow is stored in sediments above an old stonework bridge and seeps slowly into the channel below.

Ephemeral Desert Pools

Ephemeral pools sometimes form on the bajada in the Recreation Area during periods of heavy rain, especially summer thunderstorms. The few pools that last long enough for amphibian reproduction have all been in disturbed situations, where construction has altered the soil surface, creating depressions or impeding runoff. At times over the last two decades such pools have formed at the shuttle service area, Lowell Administrative Site, the Visitor Center, the shooting range, and, most recently, in detention basins south of the shooting range (see Ecological Change: Other Human Pressures).

Small artificial watering holes on the bajada, lined with concrete and fed by pipes, have sometimes been used by amphibians that naturally breed in ephemeral pools. There are currently at least two such sites: at the Visitor Center and near the road to Upper Sabino Canyon between Lowell Administrative Site and Cactus Picnic Area.

Figure 22a Temporary pool in secondary stream channel, Sabino Creek, March 2003. Lower Sabino Canyon, looking southward (downstream) toward boundary from road, camera at 2680 ft. Southern cattail, bottom center, grows among smaller emergent plants in this marshy environment. Several exotic grasses grow here, including bamboolike giant reed at left, smaller fountain grass clump at far right, rabbitfoot grass and Bermuda grass along the shore. Tufted perennial surrounding velvet ash at center is native deer grass. When free of fishes and crayfish, pools of this sort host diverse invertebrate communities. Red-spotted toad and Couch's spadefoot have bred here during the summer monsoon.

Figure 22b Temporary pool in Sabino Creek, after Aspen Fire, August 2003. Five months later the largest of multiple pulses of monsoon runoff from the Aspen Fire has rushed down the channel. Giant reed stems are pushed over and choked with pine needles, oak leaves, and bits of charcoal and wood; smaller grasses have been eroded or buried in ash-blackened sediments. A large log has fetched up against velvet ash at right, and below it, on shore of pool, a small tamarisk is partly covered with debris. Earlier in summer, Couch's spadefoot tadpoles matured and transformed here between floods, but a second cohort of tadpoles was washed away.

Ecological Change

The following chapter describes events and trends in the recent environmental history of the Sabino Canyon Recreation Area. Many of these have implications beyond the Recreation Area's amphibians and reptiles. For a list of key events in the environmental history of Sabino Creek, see Appendix D.

Drought and Fire

Persistent drought has significantly affected the Recreation Area's biotic communities in recent years. By spring 2003 a rainfall deficit had been accumulating since the late 1990s in the Recreation Area, as elsewhere in the western United States. Some communities of drought-tolerant plants had been visibly affected (figs. 3b, 11b), and activity levels of some reptiles were noticeably depressed relative to our observations two decades earlier (see gopher snake account, for example). Persistence of this drought for a decade or more, as some have predicted, would be expected to reduce some terrestrial animal populations due to diminished food availability. However, some of the strongest effects of the drought to date have been on aquatic species, through the intermediate agencies of wildfire and altered streamflow.

In the past decade several drought-related wildfires have occurred in the watersheds of the Recreation Area's mountain canyons. In 1995 the Sabino Fire burned nearly 500 acres on the southeastern slope of Upper Sabino Canyon. Effects on Sabino Creek were mostly limited to local sedimentation below side drainages and a transient algal bloom. (For more on this fire and its causes, see Nonnative Plants section.) In 2002 the 30,000-acre Bullock Fire, a major conflagration higher in the Santa Catalinas, burned only a small area of the upper Bear Creek watershed and had no discernible detrimental effects on that stream in or near the Recreation Area.

In June 2003 the Aspen Fire ignited in coniferous forest in the Pusch Ridge Wilderness. Fed by fuels accumulated during many decades of fire suppression, over the next month it spread across nearly 85,000 acres, becoming the largest and most destructive fire in the Santa Catalina Mountains in more than a century. Only small areas in and near the Recreation Area burned, but above the Recreation Area most of the Sabino Creek watershed and much of the Bear Creek watershed were within the fire's perimeter. In both streams the ensuing monsoon rains resulted in repeated pulses of black, ash-filled water, the strongest of which took the form of violent flash floods, heavily laden with inorganic sediment and organic debris (see fig. 22b). Parts of the upper Rattlesnake Creek watershed burned as well, and that stream also flooded several times.

Changing Streamflow Patterns

Preceding the dramatic events of 2003, Sabino Creek had become more prone to both high and low flow, and the changes had been most marked in the previous decade. Changes in Sabino Creek are well documented by gauge records maintained by the U.S. Geological Survey (www.usgs.gov). Changes in Bear Creek are poorly documented but appear to be similar.

Since the mid twentieth century there has been a marked increase in the intensity of the largest floods in Sabino Creek. Peak flows have been monitored since 1932, and before 1954 no floods exceeded 4000 cubic feet per second (cfs). Since 1954 there have been eight floods exceeding that level, and the greatest of these by far were the two most recent, in January 1993 (12,900 cfs) and in July 1999 (15,400 cfs).

Gauge records also indicate more prolonged periods of extreme low water (defined here as 0 cfs recorded flow) in Sabino Creek since mid century. At such times much of the channel is dry. From 1932 to 1952 the number of 0-cfs days in a calendar year never exceeded eighty. That number has exceeded eighty in half of all years since then for which there are records (there is a gap in the record for 1974–1989). The four years with the greatest recorded number of 0-cfs days have all been in the last decade, reflecting the above-mentioned drought: 1995 (167 days), 2000 (185), 2001 (161), and 2002 (176).

Effects of increased flood intensity were already clearly visible in Sabino Canyon's riparian communities before the Aspen Fire (fig. 17b). Effects of extreme low flow on the aquatic communities were amplified by sediment accumulation behind the stone bridges. At the peak of the severe foresummer drought in June 2002, there was no surface water in Upper Sabino Canyon from the fourth bridge upstream to above the eighth bridge. (That drought indirectly benefited reproduction for several native amphibians—an example of the complex and sometimes unexpected ecological effects of altered streamflow patterns. See Sonoran Desert toad, red-spotted toad, and canyon treefrog accounts.)

These trends in streamflow are likely to continue. Although the larg-est flood in summer 2003 peaked at only 3900 cfs, the potential for tor-rential flooding will persist for years, until the upstream vegetation heals. Likewise, prolonged periods of low water will recur as long as the present drought continues, and surface water will disappear even more quickly from many pools until sediment deposited in 2003 is scoured out by floods bearing lighter sediment loads. For now there has been a nearly complete loss of the larger perennial pool environments in the Recreation Area (see fig. 21b). Although the full consequences of the Aspen Fire to Sabino Can-yon's aquatic communities are not yet clear, it is already apparent that ma-jor changes have taken place. Some of these changes are mentioned below, as well as in the amphibian and reptile species accounts.

Fishes, Crayfish, and Bullfrogs

Distributions of several important aquatic predators in the Recreation Area changed markedly during the period of this study, and they may con-tinue to do so as a result both of altered streamflow patterns and of efforts to preserve a native fish. Such changes have significance for amphibians that breed in the mountain streams.

By the early 1980s only a single native fish remained in the Recreation Area, the Gila chub in Upper Sabino Canyon (native longfin dace and Gila topminnow having been extirpated at mid century). Two exotic fishes were then well established in Sabino Creek: the mosquitofish, up to the Rattle-snake Creek confluence, and the green sunfish, up to the second bridge in Upper Sabino Canyon. (We observed several other exotic fish species in Sa-bino Creek during the early 1980s, but only in Lower Sabino Canyon, near the southern Recreation Area boundary.) The mosquitofish was extirpated by the January 1993 flood, whereas the green sunfish expanded upstream, gradually replacing the Gila chub, and reached the ninth (highest) bridge by 1995.

In 1999 Sabino Creek was treated with a fish toxicant between the ninth bridge and the Sabino Lake dam, to eliminate green sunfish from that reach and allow recovery of the Gila chub population. Chub had be-gun recolonizing the treated area when the Aspen Fire ignited in 2003, and the imminent threat of flash flooding prompted biologists to remove approximately 1000 fish for holding off-site. It appeared at first that all remaining fish, both sunfish and chub, had been killed by ash-laden water after the fire, but in June 2004 a small surviving population of chub was discovered upstream from the Recreation Area boundary. (It is not known whether any sunfish survived in Sabino Creek below the National Forest boundary). In May 2005, shortly before this book went to press, several hundred of the chub removed in 2003 were reintroduced into Upper Sabino Canyon.

Fish populations in Bear Creek have been poorly documented. We observed no fish in and just above the Recreation Area in the late 1970s, but by 1981 mosquitofish, green sunfish, and largemouth bass were all present. We found only sparsely distributed green sunfish in 2002, and most of those appear to have been killed by flash flooding after the Aspen Fire, though in April 2005 we found a small number surviving below the National Forest boundary. In May 2005, eighty-five of the Gila chub removed from Sabino Creek during the Aspen Fire were reintroduced into Bear Creek, approximately 4 mi. upstream from the Recreation Area boundary.

The northern crayfish has lived in Sabino Creek since at least the 1960s, and before the Aspen Fire it was found to slightly above the Recreation Area boundary. We first observed northern crayfish in Bear Creek in 1981. This exotic crustacean has been strongly implicated in the decline of native plants and animals in Southwestern aquatic ecosystems, and it appears to have contributed to declines of several amphibians and an aquatic turtle in the Recreation Area. Sabino Creek's abundant crayfish population was decimated by the events of summer 2003, but small numbers could still be found afterward in Lower Sabino Canyon. Bear Creek's crayfish population was relatively sparse before summer 2003; it appears to have been little affected by the flooding.

The American bullfrog also arrived in Sabino Canyon by the early 1960s, and its waxing and waning numbers may have affected the native lowland leopard frog. Post–Aspen Fire flooding did not completely eliminate the very sparse population in Sabino Creek, and in late summer 2003 bullfrogs showed up in Bear Creek within the Recreation Area for the first time in our records.

For a review of the effects of exotic predators on Southwestern amphibians and reptiles, see Rosen and Schwalbe (2002).

Nonnative Plants

Since 1980 populations of some invasive exotic plants have increased significantly in the Recreation Area, and several of these have the potential to cause significant ecological change. Sweet resinbush, a perennial shrub from South Africa, expanded explosively from an old planting at Lowell Administrative Site into the foothills west of Lower Sabino Canyon. Fortunately, repeated manual removal has since controlled its spread. African sumac, a tree widely planted in the Tucson area, has proliferated in the forest at Sabino Lake (fig. 18b) and threatens to spread more widely. It is already present in riparian communities below the Recreation Area boundary and widespread along arroyos in Tucson. Tamarisk (salt cedar), a Eurasian tree that seriously infests many Southwestern riparian communities, grows today as scattered saplings along Bear and Sabino Creeks (fig. 22b), where it was rare in the early 1980s.

Exotic grasses represent a major threat to upland communities in the Recreation Area. The June 1995 Sabino Fire, carelessly ignited by a visitor at the roadside in Upper Sabino Canyon (fig. 4), was fueled in part by dry red brome, a winter-spring annual grass introduced from the Mediterranean region. The fire spread up the canyon more than a mile through the botanically diverse desert/grassland ecotone. Although wildfire during the arid foresummer can be a natural occurrence in semidesert grassland, here, as elsewhere in the Arizona Upland, this exotic grass carried the fire through paloverde-saguaro desertscrub that does not normally burn, resulting in serious damage and death to native desert plants. Many saguaros were killed; damage was especially severe directly upslope from the ignition point, where fuels were preheated by rising flames.

While red brome has grown in the Recreation Area for decades, more recently buffelgrass, a coarse African perennial, has taken hold in many locations in and near the Recreation Area and is spreading aggressively (fig. 23). Highly combustible yet fire tolerant, buffelgrass often initiates a cycle of repeated wildfires in which it eventually replaces native desert vegetation. Forest Service personnel and volunteers have begun manual removal of this grass (as well as fountain grass, Lehmann lovegrass, and natal grass, which are spreading less aggressively) from disturbed environments in the Recreation Area, but their efforts have not yet focused on the rapidly advancing invasion of buffelgrass in undisturbed plant communities.

Released Pets

Because of its easy accessibility and the unusual environment at Sabino Lake, Lower Sabino Canyon has long been used by some visitors as a place to release unwanted exotic pets. In 2002 alone we observed a large goldfish in the creek there, removed an injured painted turtle from the lake, and received an unconfirmed report of a Russian tortoise in the area. While no releases during the period of this study have resulted in the establishment of breeding populations, such remains a possibility. Moreover, a released pet could introduce a devastating exotic disease into a native animal population. (Something like this may have occurred already in the case of the

Figure 23 Buffelgrass in paloverde-saguaro desertscrub, October 2002. Looking northeastward across Sabino Canyon from bluff at old gauging station site, camera at 2740 ft. Vegetation growing between foothill paloverdes, saguaros, and other native desert plants is the exotic perennial, buffelgrass. Such dense infestations directly alter physical structures and species compositions of plant communities in ways likely to be detrimental to native animals. Moreover, spread of this fire-adapted grass threatens to start a cycle of repeated wildfires, deadly to some native animals such as desert tortoise, and resulting eventually in transformation of native desert and semidesert plant communities into unnatural grassland communities.

lowland leopard frog; see species account.) In 1984 a special order was published by Coronado National Forest prohibiting the introduction of plants and animals in the Recreation Area. Very few visitors are aware of the order, however, and the Forest Service itself has sometimes employed nonnative plants in revegetation projects.

Other Human Pressures

Visitation to the Recreation Area rose dramatically during the period of this study. Forest Service estimates indicate that the number of visitors roughly quadrupled between 1984 and 1999, from approximately 300,000 to 1,300,000 per year. The ecological consequences of this trend have not been well documented. Most recreationists are concentrated in a narrow zone on the floor of Sabino Canyon, where even conscientious visitors inevitably disturb plants and animals in the riparian and aquatic communities to an increasing degree, and inadvertently spread seeds of exotic plants. In addition, there is presumably increased pressure from the small subgroup of visitors who vandalize vegetation, harass and kill wildlife, and remove animals and plants from the Recreation Area.

Since the 1980s dense housing construction has occurred on the bajada immediately to the west and south of the Recreation Area boundaries (fig. 2). Some residents of nearby neighborhoods have reported a decrease in native wildlife, and distributions of bajada-dwelling species are being fragmented by suburban sprawl. Fortunately, both Sabino Creek and Bear Creek remain as corridors along which wildlife can move between the Recreation Area and downstream areas, and the Recreation Area remains connected to relatively undisturbed biotic communities in the protected Pusch Ridge Wilderness.

The Tucson Rod and Gun Club's shooting range, which opened in 1952, is a unique case of ecological damage in the Recreation Area. The Forest Service closed the facility in 1997, citing safety concerns related to nearby houses and schools. Later, soil containing abundant lead bullets and shot was scraped and trucked from the site, and three detention basins were excavated in drainages south of the range to trap materials washing downstream. However, lead bullets and shot still contaminate the soil on the periphery of the site. Plants immediately to the north have been damaged by stray bullets, and the core of the facility has been disfigured by large earthen berms and is largely devoid of native vegetation. The status of the shooting range is under review, but regardless of whether it reopens the ecological damage will be long lasting.

Introduction to the Species Accounts

The following annotated checklist of the herpetofauna of the Sabino Canyon Recreation Area includes nine amphibians and forty-eight reptiles, a total of fifty-seven species. Of these, six species are listed as hypothetical (their presence in the Recreation Area needs confirmation) and one species is possibly extirpated. Six species are exotic, but only one of these has yet established a breeding population in the Recreation Area. All the non-hypothetical species listed have been recorded in the Recreation Area or within a half mile of its boundaries.

The species accounts emphasize the status of populations in the Recreation Area and behavior likely to be seen by visitors. We have withheld certain information to protect sensitive species sometimes targeted by poachers, and except for brief descriptions for identification purposes we do not for the most part repeat information available in popular field guides. For more general information concerning the species listed here and for further aids to identification, the following guides are useful and easily obtained: Behler and King (1979), Smith and Brodie (1982), and Stebbins (2003). Lowe's (1964) annotated checklist summarizes the distributions of species in Arizona. The accounts follow a loose format, as explained below.

Abundance is ranked in four categories: abundant, common, uncommon, and rare. These indicate the relative frequencies with which species are usually seen in their habitats during the hours and seasons they are active. The rankings therefore refer to readily observable rather than actual abundance. This distinction is especially important in the case of secretive species, such as burrowers, which may be rarely seen despite being locally common or abundant. (The abundance categories are applied differently to the snakes than to the other groups; see the introduction to the reptile accounts.) Unless otherwise stated, abundance rankings in the accounts apply to the early 1980s; observed abundances of many species were lower in 2002 and 2003, which were marked by severe winter and spring drought.

Under Habitat and Localities we list environments and sites based on our own observations, vouchered museum records, and reliable sight reports. In some accounts these listings are followed by environments and sites in which species are expected to be found but for which we have no records.

Under Identification we emphasize those characteristics and field marks likely to be useful in distinguishing each species from others in the Recreation Area. Except where indicated, colors and patterns are those on the top of the body (dorsal), as typically seen in the field. Most amphibian call descriptions are based on Stebbins (2003). Where identification presents special problems, we include additional information in the body of the account.

Daily and seasonal activity patterns, where given in the body of the account, indicate when species are usually seen. Animals are often active but inconspicuous (underground, for example) at other times. Daily activity patterns may vary seasonally; an animal described as "diurnal and nocturnal" is likely to be primarily diurnal during the cooler seasons, but primarily nocturnal during the summer. Both diurnal and, especially, nocturnal species may show crepuscular (twilight) activity, but for brevity's sake we have usually omitted this term from the accounts.

Maps display documented distributions of species, based on our own observations and on reliable sight reports. Black symbols are for 1979–1983 and are intended to give historical snapshots of species distributions during that five-year period. Gray symbols are for later records (up to and including 2003) that add substantially to the record or indicate change. Many records have been omitted to avoid overlapping symbols. Museum records are not mapped, as most predate the map time interval and have localities too imprecise for the map scale. However, some are included in the written locality lists, which should therefore be consulted with the maps to give complete impressions of species distributions.

Finally, note that animals can and do occasionally appear at times and places outside the usual patterns outlined in the accounts. Causes of unusual activity may be evident, such as a spell of exceptionally warm winter weather, or they may not.

Amphibians

The known and hypothetical native amphibian fauna of the Sabino Canyon Recreation Area includes seven frogs and toads. One of these, the lowland leopard frog, may have been extirpated during the period of this study. Two exotic amphibians also have been observed here, the tiger salamander and the American bullfrog, though only the bullfrog has bred in the Recreation Area.

For most visitors, amphibians are an inconspicuous element in the herpetofauna. This is because, with the exception of the canyon treefrog (and very young toads and spadefoots, shortly after metamorphosis), all the common amphibians in the Recreation Area are active mostly after dark. Indeed, even the canyon treefrog is mostly inactive, though resting in the open, during daylight hours—making it, oddly, an essentially nocturnal animal that is usually seen during the day. The greatest amphibian activity occurs on warm summer nights when the ground is wet from monsoon rains. At such times amphibians may appear with an abundance rivaling that of reptiles during the day.

For the Recreation Area's toads and spadefoots, the summer monsoon is not only the time of greatest surface activity, it is also the usual breeding season. The treefrogs and frogs all begin breeding earlier, however, in either late winter or spring. The canyon treefrog and the American bullfrog go on to breed in the summer as well, but the lowland leopard frog almost invariably skips the summer and resumes breeding in the fall.

Almost all the native toads and frogs breed here primarily in streams in the mountain canyons. These aquatic environments—above all, Sabino Creek—are the key to amphibian diversity and abundance in the Recreation Area. The important exception is Couch's spadefoot, which typically breeds in summer rain pools on the bajada; but even this species has bred in Sabino Creek, in a quiet pool in an overflow channel in Lower Sabino Canyon. In fact, nearly all the frogs and toads known in the Recreation Area, including the exotic American bullfrog, not only inhabit but also have bred in Lower Sabino Canyon—a locus of exceptional amphibian diversity in the Recreation Area.

The Recreation Area's canyon streams all experienced severe flash flooding in the aftermath of the 2003 Aspen Fire, and these events altered the breeding habitats for amphibians in important ways (see Environmental Change chapter). The immediate effects of the flooding varied from species to species, as described in the following accounts.

Salamanders

Tiger Salamander
Ambystoma tigrinum

ABUNDANCE	Rarely seen
HABITAT	Perennial canyon streams, riparian scrub, riparian woodland and forest
LOCALITIES	Upper and Lower Sabino Canyon
IDENTIFICATION	The Recreation Area's only reported salamander. Yellowish, greenish, or gray, variably spotted or barred with black. Larvae superficially resemble tadpoles of frogs and toads, but retain external gills until metamorphosis.

In the Tucson area tiger salamanders are usually seen on or after wet nights during the summer monsoon. Activity is both diurnal and nocturnal.

The tiger salamander is widespread in North America but with only a scattered distribution in the arid West, where many local populations appear to be introduced. Such has certainly been the case in Sabino Canyon. Larval salamanders ("water dogs") and sometimes juveniles are sold as fish bait in Tucson, and the adults occasionally turn up in backyards. Salamanders of various ages may have been released in Sabino Creek over the years, but this species is unlikely to become established in the canyon—an arid environment in which the stream is subject to periodic severe flooding and is inhabited by large predators harmful to salamanders. Elsewhere in the Tucson region reproduction is during late winter, in ponds.

Before the Recreation Area was closed to private cars in the early 1980s, visitors occasionally captured adult tiger salamanders in and near Sabino Creek and brought them to the University of Arizona for identification, though apparently none were catalogued in the collection. We know of no sightings in recent years.

Toads and Frogs

Couch's Spadefoot (Desert Spadefoot)
Scaphiopus couchii

ABUNDANCE   Common

HABITAT   Primarily bajada desertscrub, adjacent paloverde-saguaro desertscrub; also (in Lower Sabino Canyon) riparian scrub, mesquite bosque, riparian woodland and forest. Breeds here primarily in ephemeral desert pools, occasionally also in perennial canyon stream, ephemeral desert stream (see below).

LOCALITIES   Bajada, lower margins of foothills, Lower Sabino Canyon

IDENTIFICATION   Toadlike, with widely spaced eyes; yellowish, marked with a black, brown, or greenish network (often faint in males). Pupil vertical, no clearly defined large poison gland behind eye, dark fingernail-like spade on hind foot. Call like the plaintive bleat of a lamb. (Compare with Mexican spadefoot.) Map 1, pl. 1

Couch's spadefoot is nocturnal and usually seen July to September, during and following the monsoon storms that stimulate its activity. It stays underground and inactive the rest of the year. The spades on the hind feet aid in burrowing into the soil.

This strongly desert-adapted amphibian breeds primarily in ephemeral summer rain pools, but in the Recreation Area there are few such pools that last even the remarkably short time (eight to fourteen days) needed for its tadpoles to mature and transform. At one pool we recorded the period from egg-laying to metamorphosis as nine to ten days.

We have seen successful reproduction in several rain pools on the bajada, all artificial situations: in recent years at the Visitor Center, in a soil depression filled by runoff from the roof and a landscape irrigation system; at the shooting range, in a pool next to an earthen berm; and in detention basins excavated south of the shooting range in an environmental mediation project. We have noted breeding and larvae in pipe-fed watering holes on the bajada, but we do not know if any tadpoles survived to metamorphosis.

We have also seen successful reproduction in two stream pool environments in the Recreation Area. The first is in Cholla Wash, near the southern

Recreation Area boundary (see Aquatic Environments: Ephemeral Desert Stream). The second is in Lower Sabino Canyon, in a pool that often forms in a secondary channel of Sabino Creek (figs. 22a, b). For a short time in August 2002, and again in August 2003 despite sediment- and ash-laden runoff from the Aspen Fire, tiny spadefoot metamorphs hopped at the margins of this pool among equally tiny red-spotted toads and Sonoran Desert toads dispersing from the primary stream channel.

Prior to the advent of stock ponds, channel-associated pools and ephemeral springs were likely the primary regional breeding habitat for most toads and spadefoots, and Couch's spadefoots breed regularly today in washes around urban and suburban Tucson, including farther downstream in Sabino Creek, especially where it approaches the valley floor.

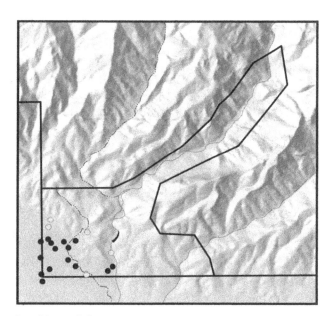

Map 1

• Couch's spadefoot

Mexican Spadefoot (Southern or New Mexican Spadefoot)
Spea multiplicata

ABUNDANCE	Hypothetical
HABITAT AND LOCALITIES	Possibly a rare inhabitant of bajada desertscrub and ephemeral desert pools, and (in Lower Sabino Canyon) riparian woodland, mesquite bosque, and riparian scrub.
IDENTIFICATION	Toadlike, brown or gray with darker blotches and many red-tipped warts. Pupil vertical, no large poison gland behind eye, dark fingernail-like spade on hind foot. Call a metallic snore. (Compare with Couch's spadefoot.)

Mexican spadefoots are usually seen during the summer monsoon in the Tucson area, where they are primarily residents of valley-floor environments, especially larger, fish-free rain ponds. They are active at night. In July 2002 we found calling male Mexican spadefoots in an artificial pond near Sabino Creek, 1 mi. south of the Recreation Area boundary, and it is therefore quite likely that individuals occasionally make their way into the Recreation Area.

Sonoran Desert Toad (Colorado River Toad)
Bufo alvarius

ABUNDANCE	Common
HABITAT	Most often seen in riparian woodland and forest, riparian scrub, mesquite bosque, nearby paloverde-saguaro desertscrub; less often at distance from aquatic and riparian environments in paloverde-saguaro desertscrub, bajada desertscrub, semidesert grassland. Breeds here in perennial canyon streams.
LOCALITIES	Upper and Lower Sabino Canyon, Bear Canyon, foothills, bajada
IDENTIFICATION	The Recreation Area's largest toad. Very large, brown or gray, sometimes with reddish warts. Pupil horizontal, large elongated poison gland behind eye, white tubercles at corner of mouth. (Toadlets are difficult to distinguish from red-spotted toad; see below.) Call a hoarse cry, rising in pitch. Map 2, fig. 24, pl. 2

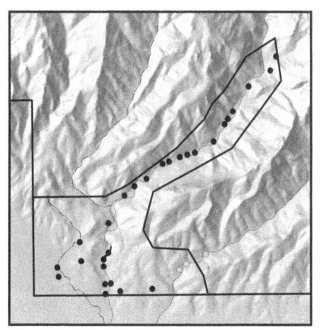

Map 2

● **Sonoran Desert toad**

Adults are mostly nocturnal, though active day and night when breeding. They are most frequently seen during the summer rains, but occasionally throughout the warm season. During the monsoon, beginning in late June or early July, large adult Sonoran Desert toads are regularly on the roads at night, mostly in wet weather, but we found no juveniles on the roads until recently (see below).

Breeding may commence just before the rains, though we have not yet observed this in the Recreation Area. Until recently, calling and amplexus (grasping of the female by the male) could usually be observed in perennial pools in Sabino Creek once the rains began, when much of the stream channel was still dry following the foresummer drought. Later breeding attempts were sometimes stimulated by the resumption of streamflow or by subsequent rain showers. Because the larger perennial pools were buried under sediments after the Aspen Fire (figs. 21a, b), reproduction may now mostly be delayed until the resumption of flow.

Female Sonoran Desert toads lay vast numbers of eggs, but we have seldom observed eggs in the Recreation Area, even in pools where amplexus has occurred. On the few occasions when we did find eggs, tadpoles

Figure 24 Sonoran Desert toads mating at night

survived for a time in abundance, but were later flushed from pools when streamflow was restored by the summer rains. When flow levels are high, tadpoles may be killed or washed downstream to below the Recreation Area boundary.

Despite these perils, in summer 2002 we found recently transformed Sonoran Desert toads among similar and much more abundant tiny red-spotted toads near the streams in both Lower Sabino Canyon and Bear Canyon. Sonoran Desert toads bred again in both streams in 2003, despite repeated flash flooding and poor water quality after the Aspen Fire. We did not observe successful reproduction in the early 1980s, and our night road hunting turned up no juveniles until late summer 2003.

Our observations suggest a population of long-lived adults that is only occasionally replenished by successful reproduction. While summer flooding is clearly one factor limiting reproduction, predation on eggs and tadpoles is likely another factor. A possibly important predator is the northern crayfish. In both 2002 and 2003 Sonoran Desert toad tadpoles survived in reaches where severe foresummer drought and predation by green sunfish had reduced local crayfish populations. (The sunfish themselves eat the tadpoles only reluctantly.)

Young Sonoran Desert toads and red-spotted toads can be difficult to tell apart. The following field marks are useful for toadlets above 0.5 in. in length.

—General coloration: Sonoran Desert toads are olive-brown, delicately marked with tiny reddish spots set in black rims. Red-spotted toads are light tan, variably and sometimes coarsely marked with reddish spots and darker pigment, usually appearing pinkish overall.

—Large poison glands: Very small Sonoran Desert toads often have streaks of yellowish pigment marking elongated glands (parotoids) developing behind eyes and orange streaks marking similar glands developing on hind legs. Red-spotted toads have concentrations of reddish spots marking nearly round parotoid glands and lack orange streaks on legs.

—Mouth warts: Before 1 in. long, Sonoran Desert toads show white swellings on dark skin at corners of mouth; by 1.5 in. length, distinctive white tubercles are clearly developed. These are absent in red-spotted toads.

See also red-spotted toad and Sonoran mud turtle accounts.

Red-spotted Toad
Bufo punctatus

ABUNDANCE	Abundant
HABITAT	Riparian woodland and forest, riparian scrub, mesquite bosque, paloverde-saguaro desertscrub, bajada desertscrub; expected in semidesert grassland. Breeds primarily in perennial and intermittent canyon streams, occasionally in ephemeral desert streams (see below).
LOCALITIES	Upper and Lower Sabino Canyon, Bear Canyon, Rattlesnake Canyon, foothills, bajada
IDENTIFICATION	Gray or tan with many small reddish warts. Flattened head, horizontal pupils, roundish to oval poison glands behind eyes. (Toadlets are difficult to distinguish from Sonoran Desert toad; see previous account.) Call a prolonged, high-pitched trill. Map 3, fig. 25

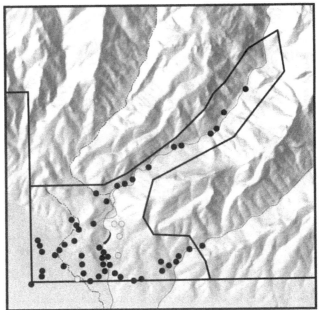

Map 3

● **Red-spotted toad**

Adult red-spotted toads are almost entirely nocturnal, even when breeding. They are usually seen April to September, but appear in large numbers only during wet weather, thus chiefly during the summer monsoon. This is the most abundant amphibian in the Recreation Area. On one rainy night in early summer 1982, before the year's crop of young had been produced, we counted ninety red-spotted toads on the roads during a single pass through the Recreation Area. After a summer's reproduction, juveniles on the roads can be almost too numerous to count.

In the Recreation Area we have seen breeding only during the summer monsoon, and almost entirely in Sabino, Bear, and Rattlesnake Creeks. Unlike the Sonoran Desert toad, which historically has bred here in quiet perennial pools at the very start of the monsoon, when the streambeds are still mostly dry, the red-spotted toad prefers to breed in flowing water. Reproductive activity for the red-spotted toad therefore begins somewhat later than for the Sonoran Desert toad, often with a noisy orgy the first night that flowing water is restored, continuing into the next morning. As a result, large numbers of transforming toads may leave Bear Creek or Sabino Creek during a brief period in August or September, sometimes covering the ground near the stream so densely that visitors must take care not to

Figure 25 Mating red-spotted toads, surrounded by freshly laid eggs

step on them. Soon afterward, dispersing juveniles may be encountered far from streams, crossing the bajada or the Phoneline Trail in Lower Sabino Canyon. A few juveniles remain active considerably later than the adults; we saw one as late as November 21. With the exception of tadpoles and (very briefly) breeding adults, these tiny animals are the only red-spotted toads likely to be seen by daytime visitors to the Recreation Area.

The reproductive head start by Sonoran Desert toads can put juvenile red-spotted toads at a distinct size disadvantage. In September 2002 we watched a young red-spotted toad being devoured by a much larger juvenile Sonoran Desert toad. However, the recent loss of large perennial pools after the Aspen Fire will likely delay breeding by Sonoran Desert toads (see account for that species).

In the early 1980s we saw little reproduction by red-spotted toads in Upper Sabino Canyon above the Rattlesnake Creek confluence. However, in summer 2002 there was a major reproductive bloom in the area of the stone bridges in Upper Sabino Canyon. An unusually deep drought had earlier dried out much of the channel there, and when streamflow resumed the toads laid eggs in water free of crayfish and fish. A similar phenomenon occurred in Bear Creek, which had experienced little flow in several previous years. In summer 2003 red-spotted toads bred in all three mountain streams, despite the flash flooding that followed the Aspen Fire.

Outside the mountain canyons, we have occasionally seen successful reproduction in persistent shallow pools in Cholla Wash, near the southern Recreation Area boundary, a site where Couch's spadefoot has also bred.

Great Plains Toad
Bufo cognatus

ABUNDANCE	Rarely seen
HABITAT AND LOCALITIES	Collected twice in Bear Canyon, in the vicinity of the former Lower Bear Picnic Area and a short distance downstream, south of the Recreation Area boundary. May be found occasionally also in Lower Sabino Canyon and on the bajada.
IDENTIFICATION	Light gray or brown with darker blotches tending to be paired around the midline, which may be marked with a light stripe. Pupils horizontal, oval poison glands behind eyes. Call very loud, a sustained clatter resembling a jackhammer.

In the Tucson area the Great Plains toad is usually seen during wet weather in the spring and summer. It is active mostly at night.

Near Tucson this toad is primarily a resident of the basin floor, but the two specimens mentioned above, both adults collected in 1968 (University of Arizona UAZ 51653, 51655), suggest that individuals occasionally move up the two perennial streams into the mouths of the mountain canyons. We have not encountered this species in the Recreation Area, though in July 2002 we found calling males in an artificial pond near Sabino Creek, 1 mi. south of the boundary. Major choruses, which are deafening, are heard mostly during the summer in the Tucson area, although calling often begins in May if water is available.

Canyon Treefrog
Hyla arenicolor

ABUNDANCE	Common
HABITAT	Riparian woodland and forest, riparian scrub; expected also in mesquite bosque and occasionally outside riparian environments in nearby paloverde-saguaro

desertscrub and semidesert grassland. Breeds in
perennial and intermittent canyon streams.

LOCALITIES   Upper and Lower Sabino Canyon, Bear Canyon,
Rattlesnake Canyon

IDENTIFICATION   Gray or tan to nearly white or pinkish, variably marked
with darker blotches. Broad pads on tips of toes. Pupils
horizontal, no enlarged poison glands behind eyes. Call
an explosive, rapid series of notes, resembling a rivet
gun.

Map 4, fig. 26, pl. 3

Canyon treefrogs are usually seen from March to November, though
sometimes during the winter months. As their name suggests, they inhabit
the Recreation Area's mountain canyons, where they are closely tied to the
rocky streams. During the day they are almost always found near water,
clinging to smooth surfaces of polished boulders and bedrock, where they
may be out of reach of garter snakes. The adult coloration, though remarkably
variable, is often an excellent match for the gneissic rock. This camouflage,
coupled with an extraordinary protective stillness—they may not move even
when touched—renders the animals nearly invisible. On wet nights some
treefrogs move well away from water, onto the roads in Sabino Canyon, and
presumably into nearby mesquite bosques. A 1975 report of "granite frogs"
in a cave above the Phoneline Trail in Upper Sabino Canyon suggests that
some canyon treefrogs travel remarkably far from streamside environments.

Breeding begins as early as March and continues into August, typically
with a break during the depth of the foresummer drought. Reproduction
is most successful in environments lacking fish and crayfish—for example,
in intermittent stream reaches and in small pools detached from the
primary stream channels, though black-necked garter snakes are often
seen preying on the tadpoles in all such places. In Sabino Creek canyon
treefrogs become abundant above what were, until the Aspen Fire, the
upstream limits of the Gila chub and the northern crayfish, just north of
the Recreation Area, and their distribution extends far upward in elevation,
into coniferous forest near the town of Summerhaven.

Treefrog abundance in the Recreation Area had already been greatly
reduced when we began our fieldwork in the early 1980s. Exotic fishes
and crayfish were likely the most important causes of the decline, but
altered streamflow patterns, sand deposition behind the bridges, and
near-constant human presence during the day may all have been factors
as well. The present distribution in Sabino Creek is reminiscent of that

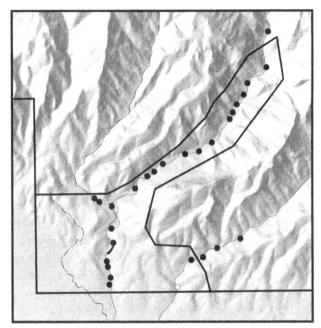

Map 4

●   Canyon treefrog

Figure 26 Canyon treefrog calling at night

of the lowland leopard frog in the 1980s, when that species existed in breeding abundance only upstream from the chub and the crayfish. However, unlike the leopard frog, the treefrog has survived and has bred in small but significant numbers within the distributions of fishes and crayfish. This difference may be attributable to the treefrog's willingness to breed in smaller pools than those favored by the leopard frog, allowing it to better take advantage of the detached, predator-free breeding habitats mentioned above.

We would rank the canyon treefrog as only uncommon in Sabino Canyon in 2002–2003. As with the red-spotted and Sonoran Desert toads, low water during the deep foresummer drought of 2002 benefited treefrog reproduction during the subsequent monsoon by reducing numbers of crayfish and fish. However, we observed no treefrog reproduction in Sabino Creek during summer 2003, in the aftermath of the Aspen Fire, though tadpoles appeared in Bear Creek in late summer, as water quality improved. By summer 2004 reproduction was recovering in Sabino Creek as well, especially in pools in the upper canyon, above the highest bridge.

The Aspen Fire's long-term effects on Sabino Creek's treefrog population will involve changes that are potentially positive (reduction of large aquatic predators) and others that are potentially negative (impaired water quality and altered streamflow). The outcome will be interesting to observe.

Lowland Leopard Frog

Rana yavapaiensis

ABUNDANCE	Possibly now extirpated in the Recreation Area and from a wide area in the Santa Catalina Mountains.
HABITAT AND LOCALITIES	Previously abundant in perennial canyon stream in Upper and Lower Sabino Canyon, and in unknown numbers in Bear Canyon. Not found in recent surveys (2001–2003) in Sabino Canyon to at least 3840-ft elevation (Hutch's Pool), nor in Bear Canyon to at least 4360 ft. (Sycamore Reservoir).
IDENTIFICATION	Long-legged and agile with a pointed snout; brownish or greenish with dark, light-rimmed spots and a light ridge on either side extending from eye toward rear leg. Call a high-pitched chuckle and a squeak like a rubbed balloon. (Compare with American bullfrog.) Map 5, fig. 27

Figure 27 Lowland leopard frog from Sabino Creek, above Recreation Area, 1981

The lowland leopard frog is active both by day and at night. It breeds in late winter through mid spring and in early fall.

A century ago this native frog was one of the most conspicuous members of Sabino Canyon's herpetofauna, and we can trace its persistence at least as far downstream as the mouths of Sabino Canyon and Bear Canyon until the late 1960s, after which no specimens from the Recreation Area were brought to herpetological collections. By the 1980s leopard frogs had apparently vanished from Bear Creek at least as far upstream as Seven Falls, but they were still abundant in Sabino Creek above 3340-ft elevation. A natural barrier at this point in the canyon, 0.4 mi. upstream from the northern Recreation Area boundary, then defined the upper limit of the Gila chub in Sabino Creek and was only 0.1 mi. above the upper limit of the exotic northern crayfish.

In the early 1980s a few leopard frogs could be found in Sabino Creek nearly down to the northern Recreation Area boundary, and in the early 1990s there were a handful of sightings in the northern part of Upper Sabino Canyon, within the Recreation Area itself. These animals are our

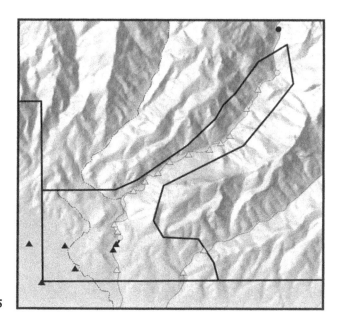

Map 5

- Lowland leopard frog
▲ American bullfrog

last records until the year 2000, when juveniles suddenly and unexpect-
edly appeared in numbers in Sabino Lake. Leopard frogs have not been
seen in Sabino Canyon since, despite surveys during 2001–2003 as high
as Hutch's Pool.

The reasons for the steep decline and apparent extirpation in the Recreation Area of this once-abundant amphibian seem to be multifaceted. In
Sabino Creek the retreat upstream of the population's lower distribution
limit closely followed the apparent first arrival of the exotic American bullfrog and northern crayfish in the early 1960s. Direct effects of the aggressively predatory bullfrog, if any, were probably mostly limited to Lower
Sabino Canyon in the 1960s, the place and time of its greatest abundance,
but the effects of the crayfish were likely more long-lasting and widespread.
This invasive crustacean tends to heavily infest habitat with few large fish,
and because both the crayfish and the Gila chub prey on tadpoles, the arrival of the crayfish left little breeding room for the leopard frog. Exotic
predatory fishes in Lower Sabino Canyon and downstream, habitat loss
due to sand deposition behind the stone bridges in Upper Sabino Canyon,

and human presence may also have contributed to the leopard frog's pre-1980 decline.

The few leopard frogs seen in the northern parts of Upper Sabino Canyon in the 1980s and early 1990s were almost certainly emigrants from the remaining population upstream. The reproductive burst in Lower Sabino Canyon in 2002 also may have originated with such emigrants, moving downstream through an environment in which green sunfish had been removed by chemical treatment in June 1999 and crayfish numbers had been reduced by the huge scouring flood a month later. However, the possibility that tadpoles or adult frogs were deliberately introduced into the lake cannot be dismissed.

The apparent demise by 2001 of the population above the Recreation Area may be due to an exotic disease (chytrid fungus disease, or chytridiomycosis) that is harming leopard frog populations through most of the Tucson region. The pathogen may have been carried into Sabino Canyon by the American bullfrog; if so, the most devastating effect of that exotic amphibian was hidden and indirect. (The introduced tiger salamander was another possible carrier.) The 2003 Aspen Fire and its consequent flash flooding were further insults to the former habitat of the lowland leopard frog in Sabino and Bear Canyons, but there is still a small chance that leopard frogs have survived somewhere in the watersheds.

American Bullfrog
Rana catesbeiana

ABUNDANCE	Uncommon
HABITAT	Riparian woodland and forest, riparian scrub. In 1980s, bajada desertscrub (see below). Has bred here in perennial canyon stream (Sabino Lake).
LOCALITIES	Lower and Upper Sabino Canyon, Bear Canyon; bajada in the 1980s
IDENTIFICATION	Long-legged, brownish or greenish, often lighter toward head, with dark-banded legs and often a spotted back. Adults heavy bodied and very large. Skin fold extending from eye behind eardrum and downward toward front leg. Call a deep-pitched bellow. (Compare with lowland leopard frog.) Map 5, fig. 28

Figure 28 Juvenile American bullfrog in rain puddle on bajada, 1980

Adult American bullfrogs are active both day and night and breed late spring to mid or late summer. Frightened young squeak when jumping into water, unlike leopard frogs.

Bullfrogs showed up in the Recreation Area about 1960, perhaps introduced for their sport or culinary values or possibly arriving on their own from artificial ponds outside the National Forest boundary. Within the Recreation Area we have seen successful reproduction only in and near Sabino Lake, the environment most like the still and slow-moving waters of its natural habitat, east of the Rocky Mountains. However, small juveniles, recently transformed from tadpoles, have sometimes moved farther up the canyon in large numbers—most recently in 1992, when we found them to above the ninth (highest) bridge in Upper Sabino Canyon—and calling adult males have occasionally been heard in the upper canyon.

This pond-dweller is poorly adapted to the torrential canyon floods of the Southwest, such as have occurred with increasing force in the Recreation Area in recent decades. The bullfrog population in Sabino Creek has responded as this situation predicts. After such events bullfrogs disappear from the Recreation Area, or nearly so; local reproduction ceases for a period of years as the stream is recolonized by fresh immigrants from outside the Recreation Area, augmented perhaps by a few flood survivors. In the early 1980s one source of immigrants was a perennial stock pond 0.4 mi. north-northwest of the Visitor Center. Juveniles dispersing from there appeared on the bajada during the monsoon rains. That pond no longer exists, but there are still bullfrog-infested artificial ponds south of the Recreation Area near Sabino Creek, an easy migration corridor. Beyond these, other potential sources remain in the Tucson Basin within the known range of bullfrog migration—more than 6 mi. overland even in Arizona.

The heyday of the American bullfrog in Sabino Canyon was in the 1960s, soon after its arrival, when it was common in Sabino Lake. The most recent reproduction in the Recreation Area was in the early 1990s. Today, after the record floods of 1993 and 1999, the population is at a low ebb. We observed only a few isolated adults and juveniles in Upper and Lower Sabino Canyon in 2002 and a single juvenile in Lower Sabino Canyon in summer 2003, during the period of flash flooding following the Aspen Fire.

Late that summer we also found several juveniles in Bear Creek, near the old Lower Bear Picnic Area. These apparent immigrants from downstream were our first bullfrog records in this reach of the stream. In the late 1970s there were bullfrogs 4 mi. upstream, in Sycamore Reservoir, but the lake has since silted in and that population no longer exists. There were still bullfrogs higher in the Bear Creek watershed, in Rose Canyon Lake, after the Aspen Fire.

Reptiles

The known and hypothetical native reptile fauna of the Sabino Canyon Recreation Area includes three turtles, sixteen lizards, and twenty-five snakes. Three additional exotic turtles have also been confirmed, and an exotic lizard is likely eventually to appear in and near buildings.

Of the three reptile groups in the Recreation Area, turtles are the least often seen, and direct human disturbance is among the causes. Shyness is a behavioral defense strategy for many turtles. After being handled or otherwise frightened they tend to retreat, hide, or sharply reduce their activity. What's more, their physical defenses are useless against people intent on removing them from their natural habitats to keep them as pets. The popularity of pet turtles has another potentially serious consequence as well: released and escaped turtles are the only nonnative reptiles that turn up regularly in the Recreation Area, and they bring with them the threat of exotic diseases.

To the Recreation Area's many daytime visitors, lizards are the most conspicuous component of the herpetofauna. Most lizards are active from March to October, but juveniles are active earlier and later in the year, when the weather is cooler, because their smaller size enables them to warm up more rapidly in the sun. Both adults and juveniles of the small side-blotched lizard are active year-round. This is usually the only reptile or amphibian seen by visitors to the Recreation Area in mid winter (though a century ago they would have found active lowland leopard frogs as well).

Despite their greater species diversity in the Recreation Area, snakes as a group are seen much less frequently than lizards. This is due to the secretive and nocturnal habits of many snakes, as well as to their generally lower actual population densities. For these reasons the scale used to rank observed abundance in the accounts for snakes is shifted relative to that used in the accounts for other groups. A rank of abundant for a snake is roughly equivalent to a rank of common for another reptile or amphibian, and other ranks on the scale are shifted correspondingly.

The reptile fauna of Sabino Canyon proper (that is, of Upper and Lower Sabino Canyon) is remarkably diverse. In addition to reptiles characteristic of rocky slopes, riparian environments, and streams in this region at

Table 2 Reptiles characteristic of higher-elevation grasslands and evergreen woodlands and forests in the Santa Catalina Mountains that also occur at or near desert elevations in Sabino Canyon.

Species	Typical Habitat at Higher Elevations	Expected or Known Habitat in the Sabino Canyon Recreation Area
Great Plains skink	grassland, woodland	riparian woodland, semidesert grassland
Madrean alligator lizard (H)	woodland, forest	riparian woodland
Graham patch-nosed snake (H)	woodland, grassland	semidesert grassland
Sonoran mountain kingsnake (H)	woodland, forest	riparian woodland
Groundsnake (H)	grassland, woodland	semidesert grassland
Ring-necked snake	woodland	riparian woodland
Arizona black rattlesnake	woodland, chaparral, forest	riparian woodland, semidesert grassland

Note: (H) indicates hypothetical species, whose presence in the Recreation Area requires confirmation.

these elevations—creatures like the eastern collared lizard, canyon spotted whiptail, and Sonoran mud turtle—the fauna is enriched by two species groups whose centers of distribution lie in other environments, outside the canyon itself.

The first group comprises lizards and snakes that in the Santa Catalina Mountains have their metropolis in higher-elevation evergreen woodlands and forests (table 2). Certain of these reptiles have taken advantage of similar ecological conditions—most obviously in vegetation structure, temperature, and soil moisture—in riparian woodland and forest communities in Sabino Canyon. As a result, their distributions extend downward to desert elevations on the canyon floor, in some cases to well below the canyon mouth.

Several of the species listed in table 2 have been reported outside riparian communities, in and near semidesert grassland on the relatively

Table 3 Pairs and triads of related reptile species associated with either mountain-canyon environments or bajada and valley-floor environments in and near the Recreation Area. All these species inhabit Lower Sabino Canyon.

Mountain Canyon	Bajada and Valley Floor
Eastern collared lizard	long-nosed leopard lizard
Greater earless lizard	zebra-tailed lizard
Clark's spiny lizard	desert spiny lizard
Sonoran spotted whiptail, canyon spotted whiptail	tiger whiptail
Sonoran whipsnake	coachwhip
Black-necked garter snake	checkered garter snake
Black-tailed rattlesnake, tiger rattle-snake	western diamondback rattlesnake

cool southeastern wall of Upper Sabino Canyon. In the Recreation Area the groundsnake and the Graham patch-nosed snake would be in the lower zones of their upland habitats, rather than in the riparian communities that primarily permit the descent of the other species.

The second group of reptiles inhabiting Sabino Canyon proper, but with distribution centers outside the canyon itself, consists of lizards and snakes characteristic of bajada and valley-bottom environments. As might be expected, the distributions of these species extend a short distance into Sabino Canyon at the canyon mouth, that is, in Lower Sabino Canyon. There, on the broad canyon floor and the lower hill slopes, they find conditions similar to their more typical habitats.

Several of these species have close relatives with very similar ecological roles (niches) in mountain canyon habitats in the Recreation Area. The result is a repeated pattern of complementary ecological distributions: one (or two) species in the mountain canyons and a different, related species on the bajada. Such species pairs and triads are listed in table 3, and where practical we have grouped them on the species distribution maps.

With infrequent exceptions, the bajada species in table 3 (and others that lack canyon counterparts, such as the side-blotched lizard) reach no farther into the canyon than Sabino Lake, which therefore marks a significant ecological boundary for Sabino Canyon's reptiles: the beginning of the true narrow canyon environment. However, the mountain canyon species all live below the lake as well as above it, with the result that all the species listed in table 3 are found together on the broad floor of Lower Sabino Canyon. For Sabino Canyon's reptiles (and amphibians), Lower Sabino Canyon is an ecotone between mountain canyon and bajada environments, and largely for this reason it supports the most diverse local herpetofauna in the Recreation Area. Yet, densities of few reptile species are notably high in this transitional environment, which has been greatly altered by recreational developments over many years, and protecting only this high-diversity zone would serve none of the species well.

Turtles

Sonoran Mud Turtle

Kinosternon sonoriense

ABUNDANCE	Uncommonly seen
HABITAT	Perennial canyon stream, riparian woodland and forest
LOCALITIES	Upper and Lower Sabino Canyon, Bear Canyon
IDENTIFICATION	Toes webbed; upper shell not highly domed, brown with three lengthwise ridges (these obscure in older turtles). Head with cream-colored stripes and mottling; white projections (barbules) on lower jaw. (Compare with pond slider and painted turtle.) Map 6, pl. 4

The Sonoran mud turtle is usually seen March to November. It is active both by day and at night.

This native turtle lives in quiet pools in Sabino Creek and Bear Creek, where it may be noticed either in the water or basking at its edge. When approached it dives to the bottom and hides among rocks, roots, or debris. Its shape and dark color, often enhanced by a light growth of turtle algae (a species restricted to aquatic turtle shells), are effective camouflage on a rocky stream bottom, but once spotted it can be easy to catch. When handled the Sonoran mud turtle emits a musky defensive odor.

The secretive behavior of this small turtle conceals a surprising abundance in some environments. A large pool in Sabino or Bear Creeks often hosts several turtles, though they may be quite invisible to visitors passing by. In 2000 a researcher trapped two dozen mud turtles in Sabino Lake—actually a

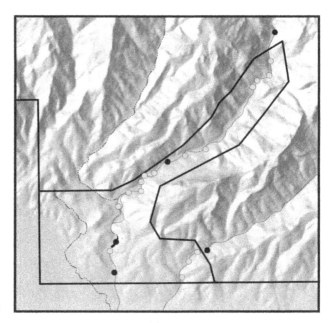

Map 6

● **Sonoran mud turtle**

sparse population for a canyon pond of this size, though much greater than most people would expect for such a seldom-seen animal.

Mud turtle sightings in the Recreation Area appear to have declined significantly since the mid twentieth century. While this might be partly due to a shift to nocturnal behavior in response to human presence, there probably has been a real reduction in population density. Possible causes include the northern crayfish (known to prey on young mud turtles), scouring floods, loss of perennial pools due to sedimentation behind the stone bridges, and illegal removal by visitors. Our observations in 2002 suggest greater abundance (certainly much greater visibility) in Sabino Creek upstream from the crayfish and the heavily visited Recreation Area—especially above the Old Dam Site (elevation 3460 ft.). Trapping in the Recreation Area in late summer 2003 showed that mud turtles survived the flash flooding that followed the Aspen Fire, and so far we have no evidence for a negative effect on them.

We once watched a young mud turtle devour large numbers of Sonoran Desert toad tadpoles with evident relish. Such predation could be a factor in the toad's limited reproductive success in the Recreation Area.

Plate 1 Mating Couch's spadefoots

Plate 2 Calling male Sonoran Desert toad

Plate 3 Canyon treefrog

Plate 4 Sonoran mud turtle

Plate 5 Male eastern collared lizard

Plate 6 Long-nosed leopard lizard

Plate 7 Male side-blotched lizard

Plate 8 Juvenile regal horned lizard

Plate 9 Canyon spotted whiptail

Plate 10 Sonoran spotted whiptail

Plate 11 Sonoran whipsnake

Plate 12 Black-necked garter snake

Plate 13 Banded sand snake

Plate 14 Juvenile western lyresnake

Plate 15 Western diamondback rattlesnake

Plate 16 Black-tailed rattlesnake

Desert Tortoise
Gopherus agassizii

ABUNDANCE	Uncommon
HABITAT	Usually seen today in paloverde-saguaro desertscrub and bajada desertscrub.
LOCALITIES	Foothills, bajada, and the three mountain canyons. Still observed, at least occasionally, in warmer environments throughout the Recreation Area.
IDENTIFICATION	Large (to 15 in. long); upper shell domed and heavy, with conspicuous growth ridges, gray or brown without contrasting color markings. Front legs heavily armed with thick scales, rear legs elephant-like. (Compare with western and eastern box turtles.) Map 7, fig. 29

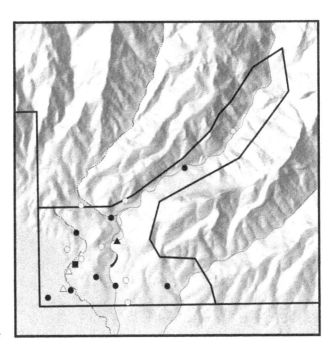

Map 7

● **Desert tortoise**

▲ **Western box turtle**

■ **Eastern box turtle**

Figure 29 Desert tortoise

Desert tortoises are usually seen March to November. They are active during the day.

Tortoises freeze at the first sight of a person, and so are easily overlooked. Typically they withdraw completely into their shells only when closely approached or physically disturbed. They should not be handled, as this can cause them to empty their bladders—critically important reservoirs for these desert animals, which reabsorb water from their stored urine. In the Recreation Area, as elsewhere, tortoises are often found with red stains around their mouths from eating cactus fruits, though their main diet is grasses and small leafy plants.

As the adjacent lands have been subdivided and developed for housing, the Recreation Area has become a remnant of comparatively undisturbed tortoise habitat at the abrupt edge of suburbia. Unfortunately, the area's tortoise population has declined over the decades, perhaps due to illegal collecting by visitors. Tortoises are now most often found away from roads and major trails. An occasional hatchling or juvenile provides evidence that reproduction is continuing, yet the long-term prospect for this population is uncertain.

The future of this charismatic reptile at Sabino Canyon will depend in large part on visitors and on residents in nearby neighborhoods where relatively sparse development could allow tortoises to persist. People should not disturb or remove tortoises; nor should they release unwanted pet tortoises, a practice that could introduce exotic diseases and incompatible genes into the population. In addition, the spread of buffelgrass (fig. 23) in the Recreation Area seriously threatens long-lived, fire-sensitive species like the saguaro and the desert tortoise, as it does entire desert communities in the Arizona Upland (see Ecological Change: Nonnative Plants). This is a growing threat, without any clear large-scale solution, although direct removal of the grass from local areas is effective.

Western Box Turtle (Ornate Box Turtle)
Terrapene ornata
and
Eastern Box Turtle
Terrapene carolina

ABUNDANCE	Rarely seen
HABITAT	Primarily riparian woodland and forest, riparian scrub, mesquite bosque, bajada desertscrub; expected occasionally in paloverde-saguaro desertscrub.
LOCALITIES	Lower and Upper Sabino Canyon, bajada
IDENTIFICATION	Toes unwebbed; upper shell strongly domed, often with a pattern of light and/or dark bars, somewhat flattened on top in western species, higher and with a central lengthwise ridge in eastern. Lower shell with a crosswise hinge, allowing front to be drawn upward. (Compare with desert tortoise, pond slider, and painted turtle.) Map 7, fig. 30

Box turtles show up occasionally both on the floor of Sabino Canyon and on the bajada. They are diurnally active.

Characteristic habitat for the western box turtle in southeastern Arizona today is level to gently rolling semidesert grassland, an environment that does not exist in the Recreation Area or its immediate vicinity. However, less than a century ago box turtles were common along the Rillito and elsewhere in Tucson's riparian bottomlands and were likely living in the woodlands along Sabino Creek, downstream from the canyon mouth and probably in the canyon itself, within the present Recreation Area. Today this original Tucson Basin population may persist largely as backyard pets, often

Figure 30 Western box turtle

carelessly hybridized with box turtles brought to Tucson from elsewhere. The animals occasionally reported in Sabino Canyon have probably been released there or escaped from nearby homes; yet, even some of these may be derived from the original Tucson Basin population. Thus, we tentatively include this turtle among the natives, rather than as an introduced species.

An adult Florida box turtle (*Terrapene carolina bauri*, a subspecies of the eastern box turtle) appeared on the bajada in September 1981. This animal, far from its natural range in the southeastern United States, is evidence that pet box turtles are indeed being released into the Recreation Area.

Pond Slider
Trachemys scripta
   and
Painted Turtle
Chrysemys picta

ABUNDANCE Rarely seen
HABITAT Perennial canyon stream, riparian woodland and forest

LOCALITIES	Sabino Lake, in Lower Sabino Canyon
IDENTIFICATION	Toes webbed, upper shell not highly domed, head with many lengthwise yellow stripes. Pond Slider often with red or orange stripe behind eye; upper shell wrinkled, usually streaked with yellow but sometimes darkening greatly with age; lower shell usually yellow with dark blotches. Painted turtle with reddish markings at edge of the smooth upper shell; lower shell reddish, usually with a dark pattern.

An adult red-eared slider (*Trachemys scripta elegans,* a subspecies of the pond slider) was seen several times in Sabino Lake during the spring of 1983. In July of that year a slider, very likely the same animal, was brought to the Visitor Center by hikers. The red-ear is the familiar dime-store turtle; hatchlings were once sold by the millions throughout the United States. They are still sold as pets today, though in much smaller numbers, and they have been widely introduced outside their natural range east of the Rocky Mountains. It is likely that many red-eared sliders have been released in Sabino Creek over the years, though evidently they do not survive for long.

Similarly, during the arid foresummer of 2002 we repeatedly observed a western painted turtle (*Chrysemys picta bellii*) in a small pool on the nearly dry bed of Sabino Lake. When we captured it, we found it had severe but well-healed injuries to its legs, including two partial amputations. (We removed this animal from the Recreation Area.) Like sliders, painted turtles are kept as pets in the Tucson area but are not native to Arizona.

Lizards

Western Banded Gecko
Coleonyx variegatus

ABUNDANCE	Uncommonly seen
HABITAT	Bajada desertscrub, paloverde-saguaro desertscrub, riparian scrub, mesquite bosque; expected also in riparian woodland and forest, semidesert grassland.
LOCALITIES	Bajada, Upper and Lower Sabino Canyon, Rattlesnake Canyon; expected also in Bear Canyon, foothills. On

Figure 31 Western banded gecko

the bajada seen especially in and near buildings. In the foothills expected especially in drainages, such as Cholla Canyon (figs. 3a, b). Known or expected throughout the Recreation Area

IDENTIFICATION   Delicate-skinned, pinkish to yellowish with brown cross-bands. Tail thickened, narrower at its base, and highly mobile; eyes with lids and vertical pupils; toes without pads. (Compare with Mediterranean gecko.) Map 8, fig. 31

This ground-dwelling gecko is usually seen late winter to early fall. However, because of its nocturnal habits, it is seldom encountered in the Recreation Area, except when accidentally uncovered in stacked wood and metal, such as at the warehouse compound. Even while surveying the roads after dark we came across a gecko only once in 1980–1983, though in July 2002 we found three on the road in Upper Sabino Canyon. Later that summer we discovered two geckos under a stone near a detention basin south of the shooting range, and one of them squeaked loudly, as banded geckos often do when disturbed. Most of our records are from pitfall traps.

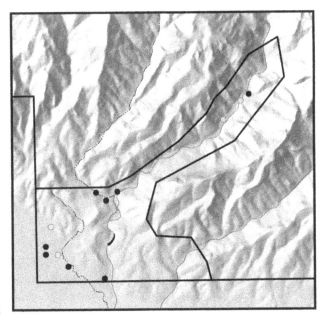

Map 8

● **Western banded gecko**

Western banded geckos turn up regularly in houses and garages around Tucson, where they are often misidentified as baby Gila monsters. However, even a hatchling Gila monster is unmistakable, with beaded skin and a broad black snout and much larger and more robust than even a full-size adult banded gecko.

Mediterranean Gecko
Hemidactylus turcicus

ABUNDANCE	Hypothetical
HABITAT	Expected eventually to appear in and near buildings in bajada desertscrub surroundings.
IDENTIFICATION	Small, skin with whitish tubercles and brown blotches; ground color changeable, from pale yellowish or pinkish to gray or brownish. Eyes lidless, with vertical pupils; toes with both pads and claws; tail thickened. (Compare with western banded gecko.)

Since the 1960s this nocturnal Old World lizard has become established in yards and around buildings in many parts of Tucson, and it is likely to show up eventually in similar habitat in the Recreation Area. The claws and toe pads facilitate its climbing habit, which readily distinguishes it from the ground-dwelling native western banded gecko. Experience elsewhere in the Tucson and Phoenix areas suggests that the Mediterranean gecko will not become widely established in the Recreation Area's natural environments, but its status should be monitored.

Eastern Collared Lizard
Crotaphytus collaris

ABUNDANCE	Uncommon
HABITAT	Paloverde-saguaro desertscrub, desertscrub/grassland ecotone, riparian scrub, mesquite bosque, riparian woodland and forest; expected in semidesert grassland.
LOCALITIES	Upper and Lower Sabino Canyon, Bear Canyon, Rattlesnake Canyon, foothills

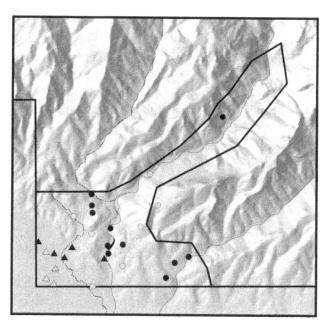

Map 9

● Eastern collared lizard

▲ Long-nosed leopard lizard

Figure 32 Female eastern collared lizard under heat stress

IDENTIFICATION Large and often colorful, with two distinct dark collars and a large head. Males bright greenish or bluish with light spots and yellowish cross-bands; females similarly patterned but less colorful overall, sometimes with orange bars on sides in mating season.
Map 9, fig. 32, pl. 5

The eastern collared lizard is day active and usually seen March to October. We have found it most often on rocky slopes, but also less frequently on the floors of the mountain canyons, including once in the riparian forest at Sabino Lake, in 1983, when that community was more open than it is today (fig. 18a).

This beautiful lizard is frequently encountered perched on a boulder in the sun. If the rock is very hot the animal assumes a stereotyped posture, lifting its body as high as possible by extending its legs full length, and minimizing its contact with the hot surface by resting on its front toes, heels, and a single point on its tail, the tip of which is raised. Similar behavior is often seen in the greater earless lizard and the zebra-tailed lizard, as well as other lizards in the Recreation Area.

Collared lizards living along well-traveled trails in the Recreation Area become habituated to humans, and they often allow very close approach. One such lizard did not retreat when one of us reached out and pulled a cactus spine from its leg.

The eastern collared lizard is replaced on the bajada by the closely related long-nosed leopard lizard (see the following account).

Long-nosed Leopard Lizard
Gambelia wislizenii

ABUNDANCE	Uncommon
HABITAT	Bajada desertscrub, mesquite bosque
LOCALITIES	Bajada; also a single sight record in Lower Sabino Canyon in 1983.
IDENTIFICATION	Large, long-tailed, grayish or tan, with an intricate pattern of round dark spots, tiny white speckles, and thin light cross-bands. Female sometimes with orange bars on sides in breeding season. Map 9, pl. 6

Like the eastern collared lizard, its close relative in rockier environments, the long-nosed leopard lizard often eats other lizards. Snakes are prey as well: a sluggish adult female we caught in September 1980 regurgitated a freshly swallowed western patch-nosed snake more than half again her length. This lizard was one of only four leopard lizards we found in the Recreation Area during our fieldwork in 1980–1983, though we have continued to see leopard lizards occasionally since then, up to the present time. Activity is diurnal.

Like collared lizards, leopard lizards are higher in the food web than other lizards and thus naturally tend to have lower population densities. Since the 1980s, housing developments on the bajada to the south and west (fig. 2) may have increasingly isolated the Recreation Area's leopard lizards, and survival of this relatively sparse population is uncertain.

Zebra-tailed Lizard
Callisaurus draconoides

ABUNDANCE   Abundant
HABITAT   Bajada desertscrub, paloverde-saguaro desertscrub,
mesquite bosque
LOCALITIES   Bajada, foothills, Lower Sabino Canyon; also recorded
twice (1958, 1968) in Bear Canyon.
IDENTIFICATION   Slender and long-legged, grayish or tan, often lemon-
yellow on sides. Tail with conspicuous black and white
cross-bands below, dark cross-bands above. Belly with
pair of black bars on either side (set in blue in male),
extending onto flanks near the middle of the body.
(Compare with greater earless lizard.)
Map 10, fig. 33

Adult zebra-tailed lizards are usually seen April to September, juveniles
February to November. Hatchlings usually first appear here in August or
September. This species is diurnally active.

In the Recreation Area the zebra-tailed lizard is primarily a resident of
desert washes and other environments with open running space, especially

Figure 33 Zebra-tailed lizard

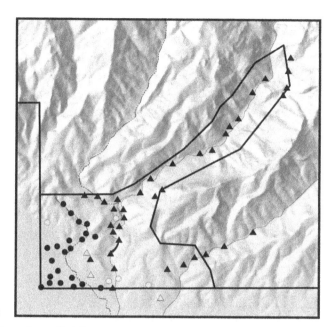

Map 10

● **Zebra-tailed lizard**

▲ **Greater earless lizard**

trails, roads, and disturbed ground near buildings. It and the closely related (and very similar) greater earless lizard are almost completely segregated in their ecological distributions—the zebra-tail on the bajada and in the foothills, and the greater earless in the mountain canyons. Two neighboring canyons provide a striking instance of this segregation. Cholla Canyon (figs. 3a, b) contains only zebra-tailed lizards, whereas Rattlesnake Canyon contains only greater earless lizards.

Along the road into Upper Sabino Canyon the transition between habitats of the two lizards is abrupt and has remained remarkably stable for more than two decades. The transition occurs at the high point in the road where it drops suddenly into the canyon east of Cactus Picnic Area. We have only a single sight record of a zebra-tailed lizard on the road east of this point (1987, south of the Rattlesnake Creek bridge). However, on the roads in the vicinity of Lower Sabino Canyon, where the edge of the mountain canyon environment is less well defined, the transition between the two lizards has been less stable, with zebra-tailed lizards sometimes appearing in the southern part of Lower Sabino Canyon and greater earless lizards sometimes in the foothills west of there.

Like Lower Sabino Canyon, the small part of Bear Canyon within the Recreation Area is primarily habitat of the greater earless lizard, though we have at least two records of zebra-tails from that locality.

For a description of this lizard's interesting tail-flagging behavior, see the following account.

Greater Earless Lizard
Cophosaurus texanus

ABUNDANCE	Abundant
HABITAT	Riparian woodland and forest, riparian scrub, mesquite bosque, paloverde-saguaro desertscrub, semidesert grassland
LOCALITIES	Upper and Lower Sabino Canyon, Bear Canyon, Rattlesnake Canyon; recorded in some years in the foothills west of Lower Sabino Canyon.
IDENTIFICATION	Slender and long-legged; grayish with small light spots, often suffused with pink, orange, or yellow in the mating season. No ear openings. Tail with conspicuous black and white cross-bands on lower surface. Belly with pair of black bars on either side (set in blue in male), often extending well upward on flanks near hind legs. (Compare with zebra-tailed lizard.) Map 10

Adult greater earless lizards are usually seen March to October, juveniles February to November. Young-of-the-year usually first appear in August and outnumber active adults by October. Activity is diurnal.

Relying largely on its speed to escape predators, this canyon-dwelling lizard is most often found in warm, rocky environments featuring flat to gently sloping ground with relatively unobstructed running space. Suitable habitat in the canyons includes, for example, expanses of sand and bedrock, roads (fig. 7), and trails (fig. 10). In the last two situations the greater earless lizard is particularly likely to be seen by visitors, who may easily mistake it for the similar zebra-tailed lizard. However, for the most part the two lizards occupy distinct environments in the Recreation Area. (See zebra-tailed lizard account.)

When approached, greater earless and zebra-tailed lizards often run briefly, then stop in full view and wave their tails, showing the banded undersurface. After repeatedly seeing this display and failing to catch the fast-running animals that perform it, a predator likely learns not to chase tail-flagging lizards. In effect, these lizards train predators not to pursue them.

The lesser earless lizard, *Holbrookia maculata*, has not been recorded in the Recreation Area. The nearest known locality is about 4 mi. southwest of the Sabino Canyon Visitor Center.

Clark's Spiny Lizard
Sceloporus clarkii

ABUNDANCE	Abundant
HABITAT	Riparian woodland and forest, mesquite bosque, riparian scrub, semidesert grassland, paloverde-saguaro desertscrub
LOCALITIES	Upper and Lower Sabino Canyon, Bear Canyon, Rattlesnake Canyon
IDENTIFICATION	Large and thick-bodied, with spine-tipped scales overlapping like shingles; gray or flecked with blue, often with irregular darker cross-bands. Black wedge on either side of neck, and dark cross-bands on forearms. (Compare with desert spiny lizard.) Map 11, see fig. 39

Adult Clark's spiny lizards are usually seen March to October. Hatchlings appear in September and remain active into November. Juveniles might be mistaken at a distance for tree lizards, which live in many of the same environments, but a young spiny lizard has a noticeably thicker body and a larger head. Activity is diurnal.

This canyon-dwelling counterpart to the desert spiny lizard is most often found in the Recreation Area in places where it can climb: large boulders, trees, rock outcrops, and masonry walls. We have seen it basking more than 30 ft. above the ground in the branches of cottonwoods. Especially in riparian environments, Clark's spiny lizard is preyed upon by a climbing snake (see Sonoran whipsnake account).

In the 1980s Clark's spiny lizards may have achieved their greatest population density at Sabino Lake, in the lakebed riparian forest and its bordering mesquite bosque (figs. 6, 18a), where these large reptiles often could be found by listening for the rustle of their scales and claws against

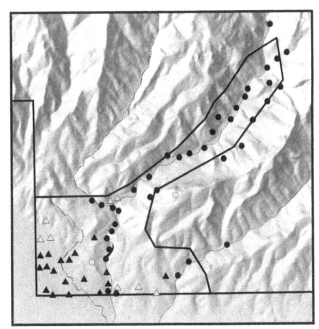

Map 11

● Clark's spiny lizard

▲ Desert spiny lizard

tree bark. Numbers may have declined at this site since then, as the forest has become denser and more shaded (fig. 18b). See the habitat discussion in the following account.

Desert Spiny Lizard
Sceloporus magister

ABUNDANCE	Common
HABITAT	Bajada desertscrub, paloverde-saguaro desertscrub, riparian scrub, mesquite bosque
LOCALITIES	Bajada, foothills, Lower Sabino Canyon, Bear Canyon in the vicinity of the former Lower Bear Picnic Area; in recent years in Upper Sabino Canyon near Rattlesnake Canyon; expected in Rattlesnake Canyon, especially where it broadens just north of the Recreation Area

Figure 34 Desert spiny lizard

boundary. Frequently seen along roadsides and around buildings.

IDENTIFICATION Large and thick-bodied, with spine-tipped scales overlapping like shingles; brownish overall, sometimes flecked with yellow on sides and with a purple patch on the back. Black wedge on either side of neck; breeding female occasionally with orange head. (Compare with Clark's spiny lizard.)
Map 11, fig. 34

The desert spiny lizard is active during the day and is usually seen March to October. Primarily a valley species, it replaces the canyon-dwelling Clark's spiny lizard on the bajada and in the foothills, where it is associated with woodrat nests, cacti, mesquite trees, and stacked lumber. It is seen routinely at the Visitor Center and is common in Tucson backyards.

The two spiny lizards are found together in Lower Sabino Canyon and in the former Lower Bear Picnic Area. On the floor of Lower Sabino Canyon we have noted a tendency for the two species to partition the environment. There the desert spiny lizard is seen mostly on the broad stream terraces south of the dam, especially among rocks where the terraces abut the floodplain, while Clark's spiny lizard predominates along the stream channel, both among boulders and in the tall trees.

The desert spiny lizard was not recorded in Upper Sabino Canyon prior to 1990, but since then we have noticed it several times along the road near the Rattlesnake Creek crossing, in narrow canyon habitat more characteristic of Clark's spiny lizard. Its recent appearance there may be related to the proximity of more congenial environments in and near Rattlesnake Canyon.

Side-blotched Lizard
Uta stansburiana

ABUNDANCE	Abundant
HABITAT	Bajada desertscrub, riparian scrub, mesquite bosque, paloverde-saguaro desertscrub
LOCALITIES	Bajada, Lower Sabino Canyon, lower borders of foothills; seen once (1979) in Rattlesnake Canyon near the Recreation Area boundary.

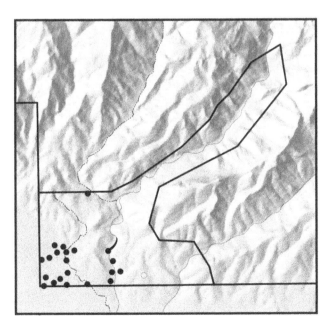

Map 12

● **Side-blotched lizard**

IDENTIFICATION   Small, brown overall, finely speckled with blue (male) or with light blotches and lengthwise stripes (female). A dark spot on either side behind front leg, often hidden by elbow in the field.
Map 12, pl. 7

The side-blotched lizard is one of the few truly winter-active lizards in the Recreation Area. It is seen year-round, though from December to February usually only on warm days and in small numbers. Hatchlings usually first appear in July. Unlike its close relative, the tree lizard, the side-blotched lizard is essentially a ground-dweller, usually found directly on the surface or perched just above it on a rock or a bit of fallen wood. Activity is diurnal.

In Lower Sabino Canyon the side-blotched lizard lives on the broad canyon floor south of Sabino Lake, on both stream terraces and sandy floodplains, in riparian scrub and mesquite bosque communities. The site in Rattlesnake Canyon where we saw it in 1979 is near an isolated patch of bajada desertscrub (see also tiger whiptail account).

Tree Lizard

Urosaurus ornatus

ABUNDANCE   Abundant
HABITAT   In all communities in the Recreation Area, but especially common in riparian woodland and forest, riparian scrub, mesquite bosque, and semidesert grassland.
LOCALITIES   In all localities in the Recreation Area, though on the bajada chiefly along washes and around buildings, and seldom seen in the foothills.
IDENTIFICATION   Small and slender, brown or gray (or nearly black when warming in the morning). Back with irregularly shaped dark bars edged with white; legs, feet and toes with numerous similar cross-bars. Usually a light upper lip visible at a distance.
Map 13, fig. 35

Figure 35 Tree lizard

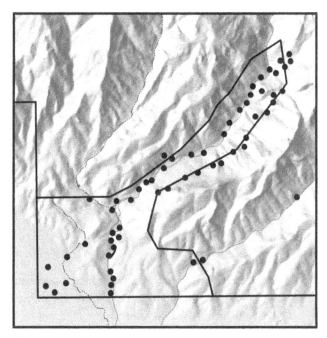

Map 13

- **Tree lizard**

Adult tree lizards are usually seen March through October, juveniles during every month except January. Hatchlings first appear in July or August. This is a day-active species.

The most abundant lizard in the Recreation Area, the tree lizard also inhabits the greatest variety of environments. A climber, it is often seen on trees, large shrubs, boulders, rock outcrops, and masonry structures; thus, in the canyons it occupies many of the same environments as the much larger Clark's spiny lizard. Mesquite trees are a favorite habitat. The tree lizard's barklike color pattern and disruptive markings make it difficult to find on tree bark unless it moves. However, like many other lizards it sometimes makes itself conspicuous by rhythmically raising and lowering its body. These push-ups are a sexual and territorial signal, usually directed at another tree lizard nearby.

On the bajada, tree lizards are often conspicuous around buildings and among stacked materials. Thoroughly pre-adapted to rough-surfaced fences and walls, these are the most common and familiar reptiles in urban and suburban Tucson.

Regal Horned Lizard
Phrynosoma solare

ABUNDANCE	Common
HABITAT	Bajada desertscrub, mesquite bosque, paloverde-saguaro desertscrub
LOCALITIES	Bajada, foothills, Upper and Lower Sabino Canyon, Bear Canyon, Rattlesnake Canyon; recorded in Sabino Canyon as far upstream as the vicinity of Rattlesnake Canyon.
IDENTIFICATION	Flattened, broad-bodied, and short-tailed; tan overall but darker along the sides. Crown of large horns at rear of the head, enlarged pointed scales scattered across the back, and a fringe of small scales along the sides. Map 14, pl. 8

The regal horned lizard is active during the day and usually seen March to October. Hatchlings are usually first seen in September.

In the Recreation Area, as elsewhere, this familiar "horny toad" feeds primarily on ants, often stationing itself near an ant trail or ant nest, where it can consume its prey in large numbers with a minimal expenditure of

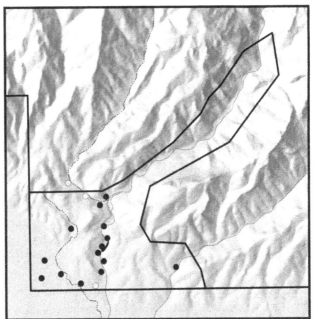

Map 14

● **Regal horned lizard**

energy. The regal horned lizard produces a neat, cylindrical dropping more than an inch long, typically composed entirely of the hard parts of a single species of ant. This large and distinctive pellet is sometimes easier to find than the highly camouflaged lizard itself.

Despite its cryptic markings, the regal horned lizard is seen fairly commonly by visitors because its preference for open ground often places it directly in their path: on trails and roadsides, in picnic areas, and near buildings. For this reason, and because this lizard is so attractive and easy to catch, the population may be adversely affected by visitors who harass or remove wildlife.

Among the many regal horned lizards we captured and measured at Sabino Canyon, only one squirted blood from its eyes—a defensive mechanism more usually directed at predators in the dog family, which dislike the taste.

Great Plains Skink
Eumeces obsoletus

ABUNDANCE	Uncommonly seen
HABITAT	Mesquite bosque, riparian scrub, riparian woodland and forest, desertscrub/grassland ecotone; expected in semidesert grassland.
LOCALITIES	Upper and Lower Sabino Canyon, Bear Canyon, Rattlesnake Canyon
IDENTIFICATION	Adults large, with a thick body and tail and short legs; smooth and shiny, greenish but often coppery on the back, marked with a fine black network like a mesh stocking. Hatchlings very different: jet black with orange and white spots on head, blue tail. Map 15, fig. 36

This secretive lizard is most often seen in the open in the spring and during the summer rains. We have captured Great Plains skinks in pitfall traps as late as September and found hatchlings in July. Activity is diurnal.

In its canyon habitat in the Recreation Area, the Great Plains skink is apparently most common in densely wooded sites, particularly among

Figure 36 Great Plains skink

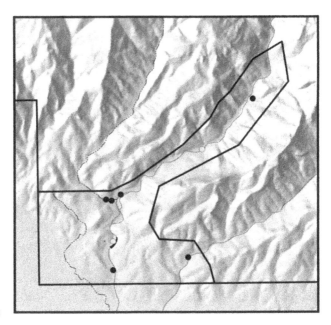

Map 15

● **Great Plains skink**

boulders where raised stream terraces border the stream channels or floodplains. This skink is seldom seen, partly because it spends much of its time under rocks and in thick vegetation, though it can occasionally be found basking in the sun. A single pitfall trap near the stream in Upper Sabino Canyon yielded four different individuals in 1981.

The Great Plains skink is one of several reptile species with populations centered at higher elevations, but extending downward into the Recreation Area on the floors of the mountain canyons (see introduction to reptile accounts).

Canyon Spotted Whiptail
Aspidoscelis (formerly *Cnemidophorus*) *burti*

ABUNDANCE Abundant

HABITAT Primarily riparian woodland and forest, riparian scrub, mesquite bosque; less frequent in semidesert grassland and less common still in paloverde-saguaro desertscrub.

LOCALITIES   Upper and Lower Sabino Canyon, Bear Canyon, Rattlesnake Canyon; also recorded once in Cholla Canyon and once in the foothills east of Bear Canyon.

IDENTIFICATION   Large; back dark with light spots between light lengthwise stripes, but these stripes becoming discontinuous in largest individuals. Tail very long and whiplike, often reddish toward tip. (Small individuals difficult to distinguish from Sonoran spotted whiptail; see below.)
Map 16, pl. 9

The canyon spotted whiptail is usually seen March to October, in the mountain canyons as befits its name. We observed mating behavior, a male closely following a female, in May. Hatchlings usually first appear in July. Activity is diurnal.

This is a bisexual species (that is, males are present; see Sonoran spotted whiptail account). The subspecies called the giant spotted whiptail, *Aspidoscelis burti stictogramma*, lives in the Recreation Area. Large adults may approach 17 in. in total length and are seen mostly April–September, during the hot core period of activity for the species. These animals are

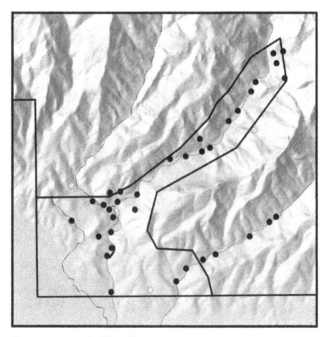

Map 16

● **Canyon spotted whiptail**

easily distinguished from Sonoran spotted whiptails by their size, but smaller spotted whiptails are more difficult to identify. The following field marks, taken in combination, help to distinguish subadult and juvenile giant spotted whiptails from similar-size Sonoran spotted whiptails.

—Tail: Brownish tan to greenish in Sonoran spotted whiptails; orange-tan to reddish in giant spotted whiptails.

—Stripes: Pair of stripes in middle of back very often pinched together toward rear of body in Sonoran spotted whiptails; roughly parallel in giant spotted whiptails.

—Spots: Seldom present between central pair of stripes in Sonoran spotted whiptails; usually present between these stripes in giant spotted whiptails; smaller and more numerous overall in giant spotted whiptails.

—Ground color: Rich brown in adult Sonoran spotted whiptails; darker and nearly black in similar-size giant spotted whiptails and in smaller individuals of both kinds.

In large adult giant spotted whiptails the stripes break up into rows of spots, yielding a coarsely spotted pattern on the back and sides, and a reddish tint develops on the nape. In the 1980s individuals with this full adult coloration could be found most regularly in the willow forest at Sabino Lake (fig. 18a), where the broad tracks left by their robust tails were a common sight in the sand under the trees. As the lakebed forest has become more dense (fig. 18b), these lizards have become more difficult to see. Perhaps the population there has declined in the more heavily shaded, cooler environment. Today, large adult giant spotted whiptails can still be found readily in the woodlands and bosques upstream from the lake into the vicinity of the Rattlesnake Creek confluence (fig. 5). They are seen only infrequently south of Sabino Lake.

Sonoran Spotted Whiptail
Aspidoscelis (formerly *Cnemidophorus*) *sonorae*

ABUNDANCE   Abundant

HABITAT   Primarily riparian woodland and forest, riparian scrub, mesquite bosque, semidesert grassland; less common in paloverde-saguaro desertscrub, and only occasionally in bajada desertscrub.

LOCALITIES   Upper and Lower Sabino Canyon, Bear Canyon, Rattlesnake Canyon; less common in the foothills bordering the lower reaches of the two major canyons, including in Cholla Canyon. Since 1986, seen occasionally on the northeastern bajada near Cactus Picnic Area.

IDENTIFICATION Chocolate brown, the back with light lengthwise stripes and often vaguely defined light spots; tail long and whiplike, often greenish toward tip. For aid in distinguishing from the canyon spotted whiptail, which inhabits many of the same localities, see account for that species.
Map 17, pl. 10

Sonoran spotted whiptails are usually seen March to October. They are day active. Captive individuals from the Recreation Area have laid eggs in May, June, and August. Hatchlings usually first appear in July or August.

What we are here calling the "Sonoran spotted whiptail" represents an important unsolved puzzle concerning the Recreation Area's herpetofauna. Since the 1960s several all-female species of whiptail lizards have been described in the scientific literature, including the Sonoran spotted whip-

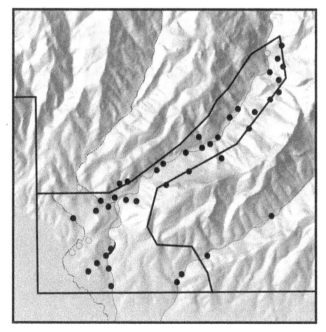

Map 17

● **Sonoran spotted whiptail**

tail, *Cnemidophorus sonorae* (Lowe and Wright 1964). Such parthenogenetic forms (laying fertile eggs without mating) have arisen repeatedly in the Southwest by hybridization between bisexual, conventionally reproducing whiptail species. Most are only a few hundred to a few thousand years old—much younger than their bisexual progenitors. Often the canyon spotted whiptail is one of these progenitors, and its rehybridization with the initial hybrid cross can generate a great diversity of unisexual lineages (each of which may be considered a species, and most of which have three sets of chromosomes, rather than the original two).

For many years unisexual lizards in the Sabino Canyon area have been referred to the Sonoran spotted whiptail species or occasionally to the Gila spotted whiptail, *Aspidoscelis flagellicauda*. However some or all such lizards in the Recreation Area may belong to a different, as-yet-undescribed lineage or perhaps to more than one such lineage. Genetic analysis likely will be needed to sort out this confusing situation.

Tiger Whiptail (Western Whiptail)
Aspidoscelis (formerly *Cnemidophorus*) *tigris*

ABUNDANCE	Abundant
HABITAT	Bajada desertscrub, paloverde-saguaro desertscrub, mesquite bosque, riparian scrub
LOCALITIES	Bajada, Lower Sabino Canyon, Bear Canyon in the vicinity of the former Lower Bear Picnic Area, lower borders of the foothills, including in Cholla Canyon; also in Rattlesnake Canyon where it broadens just north of the Recreation Area boundary.
IDENTIFICATION	Brown or tan overall, but blackened on throat and chest; light lengthwise stripes on back (obscure in adults), conspicuous light spots on back, sides, and hind legs; tail long and whiplike, in hatchlings light blue. (Compare with canyon spotted and Sonoran spotted whiptails.) Map 18

Adult tiger whiptails are usually seen April to September, juveniles March to October. Activity is diurnal. We have observed mating behavior (males closely following females, copulation), in April and May. A captive from the Recreation Area laid eggs in June. Hatchlings usually first appear in July.

This is a bisexual species. Elsewhere in its geographic range, the tiger whiptail has occasionally hybridized with the all-female Sonoran spotted whiptail, producing sterile offspring with four sets of chromosomes.

Much like the side-blotched lizard, the tiger whiptail lives primarily on the bajada in the Recreation Area and occupies only warmer, more open environments in the lower canyons. For example, in the 1980s we found this species in the disturbed mesquite bosque along the west side of Sabino Lake, but it avoided the shady willow forest of the lakebed itself (fig. 6). Canyon spotted and Sonoran spotted whiptails inhabit both communities. The area where the tiger whiptail is found in Rattlesnake Canyon has been mapped as a disjunct patch of bajada desertscrub (Lazaroff 1993: 6–7).

Unlike most lizards in the Recreation Area, which sit still and wait for prey to approach, whiptails almost constantly move about, actively hunting, pausing only now and then to warm up or cool off in a patch of sun or shade. This habit makes whiptails as a group easy to identify at a distance— as well as conspicuous prey for greater roadrunners. We once saw a tiger whiptail raise its tail, draw its forelimbs against its body, and lope (not run) across the ground on its rear legs. The function of this very unusual gait is unknown, though perhaps the lizard was tracking a flying insect.

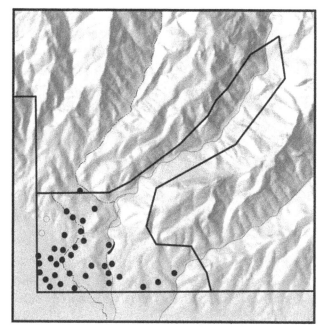

Map 18

• **Tiger whiptail**

Madrean Alligator Lizard
Elgaria kingii

ABUNDANCE	Hypothetical
HABITAT AND LOCALITIES	Expected in riparian woodland and forest, riparian scrub, and mesquite bosque in Upper and Lower Sabino Canyon and Bear Canyon.
IDENTIFICATION	Slender, small-legged, and long-tailed, pale with brownish cross-bands. Skin with square scales and a lengthwise fold or groove on either side of the body.

The Madrean alligator lizard is diurnally active.

This characteristic inhabitant of mountain evergreen forests and woodlands is seen also at lower elevations along major riparian corridors in grassland and desert surroundings, including in bottomland riparian and cienega communities in some southern Arizona valleys and basins. The species is found today near a spring on the bajada of the eastern Tucson Basin, and it may once have been more widespread near Tucson. The ring-necked snake and the Great Plains skink also have distributions centered at higher elevations, extending downward into the desert in riparian communities; both (especially the snake) are sometimes found in the same localities as the Madrean alligator lizard. The presence of these two other species in the Recreation Area supports the expectation that the Madrean alligator lizard will eventually be found there as well.

Gila Monster
Heloderma suspectum

ABUNDANCE	Uncommon
HABITAT	Bajada desertscrub, paloverde-saguaro desertscrub, mesquite bosque, riparian scrub, riparian woodland and forest. Seen once in the desertscrub/grassland ecotone in Upper Sabino Canyon.
LOCALITIES	Bajada, foothills, Upper and Lower Sabino Canyon, Bear Canyon, and Rattlesnake Canyon
IDENTIFICATION	Very large and heavy-bodied, with a conspicuously thick tail. Scales resembling beadwork, in a coarse, contrasting network of black against pink or orange. Snout black. Map 19, fig. 37

Figure 37 Gila monster

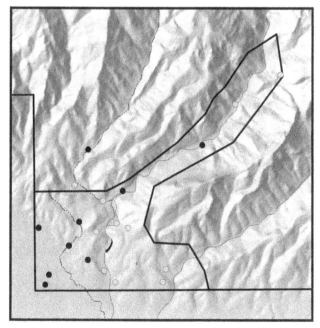

Map 19

• **Gila monster**

Gila monsters are chiefly diurnal, but juveniles and occasionally adults are active after dark in warm weather. They are usually seen March to September.

Despite their venom, Gila monsters pose minimal threat to Recreation Area visitors, because these lizards almost always retreat when approached. If cornered, they may hiss threateningly, and they can quickly strike sideways and toward the rear, but they are not known to bite unless provoked.

We observed a Gila monster in a face-to-face confrontation with an adult rock squirrel on the road in Upper Sabino Canyon. The lizard was backing up slowly as the squirrel repeatedly lunged toward it. In this canyon environment, nestling rock squirrels are part of the Gila monster's diet.

In the Recreation Area Gila monsters are by no means restricted to desertscrub communities. We have numerous sight records in and near riparian communities, including dense riparian woodland, and we have twice watched these lizards wade and swim through shallow water while crossing Sabino Creek.

Snakes

Western Blind Snake
Leptotyphlops humilis

ABUNDANCE	Rarely seen
HABITAT AND LOCALITIES	Recorded in paloverde-saguaro desertscrub near the floor of Rattlesnake Canyon. Expected in all biotic communities and localities in the Recreation Area, especially in drainage bottoms.
IDENTIFICATION	Small and wormlike, blunt on both ends, brownish to shiny pink. Eyes vestigial, reduced to dark spots; short spine on tip of tail. Map 20, fig. 38

More secretive than rare, this unusual snake spends almost all its time underground and in the nests of ants and termites, its most important prey. It occasionally appears on the surface at night. We found a western blind snake only once during our fieldwork, in a pitfall trap in April 1981.

Figure 38 Western blind snake

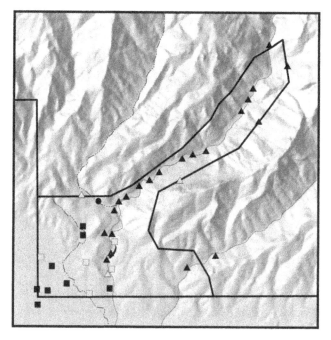

Map 20

- ● **Western blind snake**
- ▲ **Sonoran whipsnake**
- ■ **Coachwhip**

Sonoran Whipsnake
Masticophis bilineatus

ABUNDANCE   Abundant
HABITAT   Riparian woodland and forest, mesquite bosque, riparian scrub, semidesert grassland, paloverde-saguaro desertscrub
LOCALITIES   Upper and Lower Sabino Canyon, Bear Canyon, Rattlesnake Canyon
IDENTIFICATION   Agile and fast-moving, slender with a tapering, whiplike tail; greenish or bluish with lengthwise light and dark stripes along the sides, colors and markings fading toward the tail.
Map 20, fig. 39, pl. 11

This is one of the most frequently encountered snakes in the Recreation Area. It is diurnally active and usually seen March to November.

The Sonoran whipsnake is primarily a resident of riparian communities in the Recreation Area, and it replaces the closely related coachwhip in narrow canyon environments. An efficient and frequent climber, this predator is sometimes seen high in the trees of the riparian woodlands,

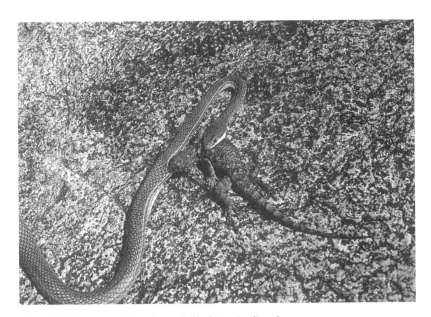

Figure 39 Sonoran whipsnake and Clark's spiny lizard

where the sounds of agitated birds may signal its presence. We have seen a Sonoran whipsnake being dive-bombed by a pair of Cassin's kingbirds 20 ft. above the ground in a velvet ash and another harassed by a pair of hooded orioles nearly twice this high in an Arizona sycamore. We also found a whipsnake in a staghorn cholla with its head in a cactus wren nest, while the adult birds fussed ineffectually nearby. By the time the snake withdrew, it had partly swallowed a well-feathered nestling.

Reptiles are also important prey for this snake. We watched a Sonoran whipsnake persistently hunting an adult Clark's spiny lizard in a mesquite. When it located the lizard on the side of a horizontal branch, it took time adjusting its purchase on the branch before lunging at the lizard, but it missed and fell anyway. On another occasion we found a whipsnake on a large boulder with its mouth clamped behind the forelimbs of a Clark's spiny lizard, which in turn had gripped the side of the snake in its jaws (fig. 39). Whipsnakes are also known to eat other snakes, including young rattlesnakes.

Mammals are taken as well. We once observed a Sonoran whipsnake swallowing a loudly squeaking rodent, apparently a very young white-throated woodrat. See also the following account.

Coachwhip
Masticophis flagellum

ABUNDANCE	Common
HABITAT	Bajada desertscrub, paloverde-saguaro desertscrub, and (in Lower Sabino Canyon) riparian scrub, mesquite bosque; expected also in riparian woodland and forest near the mouths of Sabino Canyon and Bear Canyon.
LOCALITIES	Bajada, foothills, Lower Sabino Canyon; expected in Bear Canyon in the vicinity of the former Lower Bear Picnic Area.
IDENTIFICATION	Agile and fast-moving, slender with a tapering, whiplike tail. Two color phases, with intermediates sometimes seen: (1) all black above, and, much less frequently, (2) reddish, with darker cross-bands on the neck. Map 20

This day-active snake is usually seen March to October, rarely in late winter.

The coachwhip is the Sonoran whipsnake's counterpart on the bajada and in the foothills. We have usually found this species on the ground, though it readily climbs into cacti and the small trees that grow in its desertscrub habitat, to sleep and to prey on bird eggs and nestlings. Like the Sonoran whipsnake, when approached the coachwhip usually retreats at high speed, and if caught it usually bites defensively, though neither snake is venomous. Both species sometimes vibrate their tails when disturbed, as do many other snakes.

Coachwhips feed most heavily on reptiles, both lizards and snakes, and they are strong enough to take on sizeable rattlesnakes. They are often seen near the Visitor Center in the spring, when newborn round-tailed ground squirrels are present in large numbers. The piping call of the squirrels sometimes signals the presence of a snake, which may attempt to chase down young squirrels surprised away from their burrows. Feces of a coachwhip we captured on the bajada contained lizard scales and claws, fur, and insect parts.

The red racer subspecies, *Masticophis flagellum piceus*, lives in the Recreation Area, and its mostly black color phase, sometimes called the "black racer," is most often seen. Perhaps the black coloration aids in concealment in the Arizona Upland desert environment, where the glaring sun sometimes makes it difficult to distinguish a dark snake from the shadow of a tree branch or a cactus.

Saddled Leaf-nosed Snake
Phyllorhynchus browni

ABUNDANCE	Rarely seen
HABITAT	Bajada desertscrub; expected in paloverde-saguaro desertscrub.
LOCALITIES	Bajada; expected in the foothills, and occasionally on the floors of Lower Sabino Canyon and of Bear Canyon in the vicinity of the former Lower Bear Picnic Area.
IDENTIFICATION	Cream-colored or pinkish, with brown, black-rimmed blotches draped across the back like saddles and an enlarged, raised scale looking like a thick leaf laid over the snout. Map 21

In the Recreation Area and its vicinity, the saddled leaf-nosed snake has been observed primarily in May and June. There are numerous records on Sabino Canyon Road south of the Recreation Area, but we encountered a

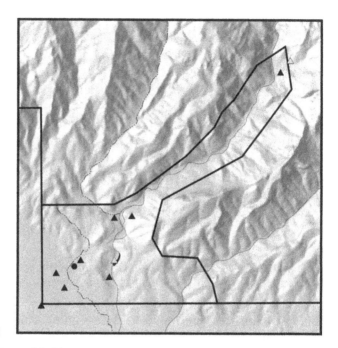

Map 21

● **Saddled leaf-nosed snake**

▲ **Western patch-nosed snake**

leaf-nosed snake only once during our fieldwork in 1980–1983 and again once in 2002–2003. Activity is nocturnal.

This is a small subtropical snake that is successful in burrowing in the coarse sandy-gravelly soils of the paloverde-saguaro upland desert. The enlarged scale on the snout would appear to be a useful structure for burrowing and for excavating lizard eggs, its primary food.

Western Patch-nosed Snake
Salvadora hexalepis

ABUNDANCE	Common
HABITAT	Bajada desertscrub, mesquite bosque, riparian woodland and forest, paloverde-saguaro desertscrub, desertscrub/grassland ecotone; expected also in semidesert grassland, riparian scrub.
LOCALITIES	Bajada, Upper and Lower Sabino Canyon; expected in

Figure 40 Western patch-nosed snake

foothills, Bear Canyon, Rattlesnake Canyon. Known or expected throughout the Recreation Area.

IDENTIFICATION Grayish, with a broad, pale, sometimes yellowish central stripe bordered by dark stripes—one on either side near the head, two on either side farther down the body. Enlarged patchlike scale on the snout. (Compare with Graham patch-nosed snake.)
Map 21, fig. 40

The western patch-nosed snake is usually seen March to November, but more often than most other snakes during the cooler seasons. It is diurnally active.

During our fieldwork in the early 1980s, we found this small, mostly ground-dwelling snake most frequently on the roads in Upper Sabino Canyon and on the bajada. Sightings were particularly common near the Rattlesnake Canyon confluence. One of our bajada specimens was regurgitated by a leopard lizard.

We did not encounter any patch-nosed snakes in 2002, when their activity may have been greatly reduced due to severe drought conditions, but we found one at Sabino Lake during the summer 2003 monsoon.

Graham Patch-nosed Snake (Mountain Patch-nosed Snake)
Salvadora grahamiae

ABUNDANCE	Hypothetical
HABITAT AND LOCALITIES	Expected in semidesert grassland in Upper Sabino Canyon.
IDENTIFICATION	Grayish with a broad pale central stripe, this stripe noticeably lighter than the sides of the body, sometimes tinted orange or yellow, and bordered on either side by a single dark stripe. Enlarged patchlike scale on the snout. (Compare with western patch-nosed snake; see also below.)

The Graham patch-nosed snake inhabits woodlands and grasslands at higher elevations in the Santa Catalina Mountains, where there are records down to about 4000-ft. elevation, but it has not yet been definitely recorded in the Recreation Area. The locality for a 1954 specimen (University of Arizona UAZ 41239) is given as "Upper Sabino Canyon," but this could be farther up, north of the Recreation Area boundary.

Any patch-nosed snake found in Upper Sabino Canyon, especially in semidesert grassland, should be closely observed in case it proves to be this species, rather than the western patch-nose. Positive identification may require examining the arrangement and number of scales on the head (Stebbins 2003). One comparison is to count the scales on the upper lip (upper labials) from the central patch scale back to the rear of the mouth. There are eight on either side on the Graham patch-nose and nine on the western patch-nose.

Glossy Snake
Arizona elegans

ABUNDANCE	Rarely seen
HABITAT	Bajada desertscrub
LOCALITIES	Bajada
IDENTIFICATION	Slender and with a narrow head; glossy and pale with a low-contrast, faded appearance, marked with many brownish or gray blotches, each outlined with black. Map 22

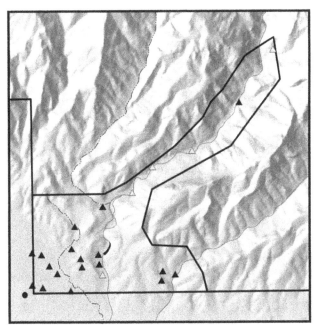

Map 22

● Glossy snake

▲ Gopher snake

Mice and lizards are favorite foods of this strong constrictor, which captures many such prey in their underground retreats. In fact, the glossy snake spends much of its time in small mammal burrows, and is active on the surface mostly at night. Records near the Recreation Area, including many to the south along Sabino Canyon Road, indicate that it is seen most frequently from April through July, prior to the summer rains. There are no definitely on-site records; the nearest precise locality is 0.2 mi. west of the Sabino Canyon Visitor Center.

Gopher Snake
Pituophis catenifer

ABUNDANCE   Abundant

HABITAT   Bajada desertscrub, paloverde-saguaro desertscrub, riparian scrub, riparian woodland and forest, mesquite bosque; expected also in semidesert grassland.

LOCALITIES   Bajada, foothills, Upper and Lower Sabino Canyon, Bear Canyon; expected also in Rattlesnake Canyon. Known or expected throughout the Recreation Area.

IDENTIFICATION   Large, tan or cream-colored, strongly marked with a row of black-edged brown blotches on the back and more numerous smaller but similar blotches on the sides. Top of head tan, speckled or spotted with black. Map 22, fig. 41

Gopher snakes are active both by day and at night and are seen year-round, though mostly juveniles in the winter. Mating has been reported in the Recreation Area in May, and hatchlings usually first appear in September. Mostly we have found gopher snakes on the ground in the Recreation Area, but we saw one shimmying up a velvet ash, and they are known to prey on bird nestlings and eggs.

Although usually docile when caught and handled, a gopher snake (often called a "bull snake") can be aggressive if grabbed or cornered. Following a nighttime confrontation with a gray fox, one snake coiled, vibrated its tail, repeatedly produced a rasping hiss resembling the rattle of a rattlesnake, and struck at one of us. This common behavior, coupled

Figure 41 Gopher snake

with the snake's vaguely diamond-shaped markings, sometimes causes a gopher snake to be mistaken for a rattlesnake and killed by visitors. Another agitated gopher snake pursued one of us several yards across the bajada during the daytime—unusually aggressive behavior for this species.

In the early 1980s gopher snakes were outnumbered only by black-necked garter snakes in our sightings, but we encountered only three gopher snakes in the Recreation Area in 2001–2003. This change could represent a true decline in population or reduced activity due to drought.

Common Kingsnake
Lampropeltis getula

ABUNDANCE	Uncommon
HABITAT	Riparian woodland and forest, riparian scrub, mesquite bosque, bajada desertscrub, paloverde-saguaro desertscrub; expected also in semidesert grassland.
LOCALITIES	Upper and Lower Sabino Canyon, bajada; expected in Bear Canyon, Rattlesnake Canyon, foothills. Known or expected throughout the Recreation Area.
IDENTIFICATION	Glossy black with whitish or yellowish cross-bands that widen toward the belly, and light bars on the snout. Map 23

Common kingsnakes are usually seen spring to fall; most of our records are April–August. Activity is both diurnal and nocturnal. The subspecies called the California kingsnake (*Lampropeltis getula californiae*) inhabits the Recreation Area.

Here, as elsewhere in the Sonoran Desert, the common kingsnake is primarily a riparian species (found near streams) or a xeroriparian species (found along dry washes). It may be encountered away from such drainage features, however, and individuals show up occasionally in and beneath buildings.

A common kingsnake was observed constricting a tiger rattlesnake in Upper Sabino Canyon. Snakes, including rattlesnakes, make up a significant part of its diet, which also includes rodents and lizards. It is resistant to rattlesnake venom, as are some other snake-eating snakes.

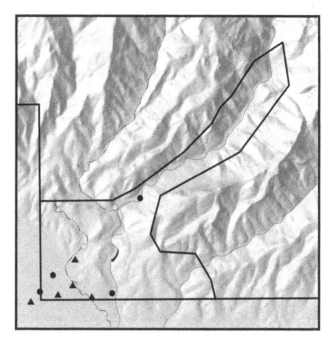

Map 23

● **Common kingsnake**

▲ **Long-nosed snake**

Sonoran Mountain Kingsnake
Lampropeltis pyromelana

ABUNDANCE	Hypothetical
HABITAT AND LOCALITIES	Possibly a rare inhabitant of Upper Sabino Canyon, primarily in riparian woodland and forest.
IDENTIFICATION	Vivid red, black, and whitish cross-bands with irregular edges, the red bands bordered by black. Snout whitish. (Compare with long-nosed snake and Sonoran coralsnake.)

In Arizona this colorful diurnal snake lives primarily in riparian communities above 4700-ft. elevation, in evergreen woodland and forest surroundings, but it sometimes descends to lower elevations in mountain canyons. In the late 1970s and early 1980s there were several unconfirmed reports of Sonoran mountain kingsnakes in Upper Sabino Canyon, and in 2002 we received a report of one along the Sabino Canyon Trail, between the

Recreation Area boundary and Sabino Basin. A specimen or photograph is needed to document the presence of this species in the Recreation Area. See also the following account.

Long-nosed Snake
Rhinocheilus lecontei

ABUNDANCE	Uncommonly seen
HABITAT	Bajada desertscrub, less commonly paloverde-saguaro desertscrub; expected occasionally (in the lower canyons) in mesquite bosque, riparian scrub, riparian woodland and forest.
LOCALITIES	Bajada, less commonly foothills; recorded once in Bear Canyon in the vicinity of the former Lower Bear Picnic Area; expected also in Lower Sabino Canyon.
IDENTIFICATION	Back marked with large black saddles, these conspicuously speckled with white on the sides; coloration between saddles highly variable, with or without red pigment and sometimes with black speckles on the sides. (Compare with Sonoran mountain kingsnake and Sonoran coralsnake.) Map 23

Activity is nocturnal, frequently crepuscular, and rarely extends into late morning. Because the long-nosed snake usually remains underground by day, it is seldom observed by visitors to the Recreation Area. It has been recorded on the roads April to October, most frequently in April and May and during the summer rains.

It is possible that some unconfirmed sight reports of Sonoran mountain kingsnakes in Sabino Canyon actually refer to misidentified long-nosed snakes. If so, the distribution of the long-nosed snake in the Recreation Area would include narrower mountain canyon environments than the localities listed above, though these would not be its usual habitat. Further documented observations are needed to clarify the status of these two snakes in the Recreation Area.

Black-necked Garter Snake
Thamnophis cyrtopsis

ABUNDANCE   Abundant
HABITAT   Perennial canyon streams, intermittent canyon stream, riparian woodland and forest, riparian scrub, mesquite bosque; occasionally in desertscrub communities.
LOCALITIES   Upper and Lower Sabino Canyon, Bear Canyon, Rattlesnake Canyon; occasionally foothills and bajada.
IDENTIFICATION   Highly aquatic. A yellowish stripe down the back (somewhat orange toward the head) and a light stripe on either side; inconspicuous dark spots between the stripes. Pair of large black blotches on neck, one on either side. (Compare with checkered garter snake.)
· Map 24, pl. 12

The black-necked garter snake is usually seen March to November. We have found newborns in July and August. Activity is primarily diurnal.

This is the most frequently seen snake in the Recreation Area. One July morning in 1980, when a front of water moved slowly down the dry bed of Sabino Creek following heavy monsoon rains, ten black-necked garter snakes of various sizes were flushed from beneath rocks as the water filled a single pool in Upper Sabino Canyon.

In the Recreation Area black-necked garter snakes are almost always found in or near rocky canyon streams. Graceful swimmers, they are important predators on canyon treefrog and red-spotted toad tadpoles, and we have seen them chasing small fish trapped in isolated pools at times of low water. Among our many sight records in the Recreation Area, only three, all of them adults, are outside the mountain canyons: two in the foothills east and west of Lower Sabino Canyon and one at the Visitor Center. (The last animal may have come from an off-site stock pond where we observed garter snakes, rather than from Sabino Creek. See American bullfrog account.) There is also a 1916 specimen (American Museum of Natural History AMNH 20591) taken near the location of the present Lowell Administrative Site. The July and September dates of these unusual records suggest that black-necked garter snakes may be most likely to leave their usual riparian habitat here during the summer monsoon.

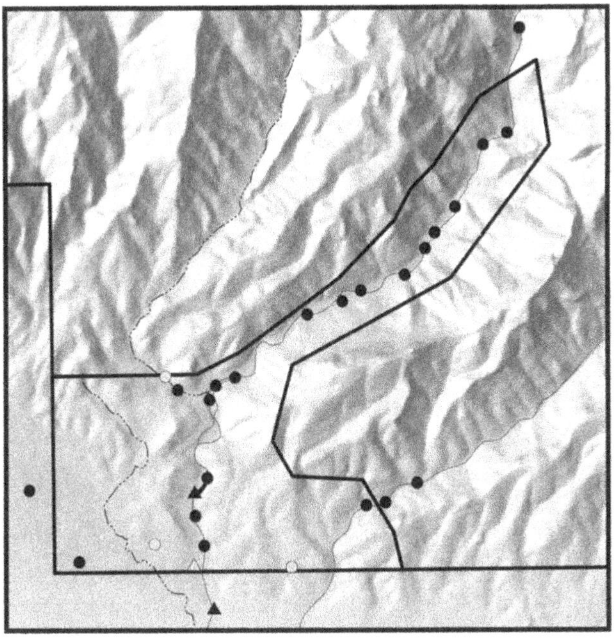

Map 24

● Black-necked garter snake

▲ Checkered garter snake

Checkered Garter Snake
Thamnophis marcianus

ABUNDANCE   Uncommon
HABITAT   Perennial canyon streams, riparian woodland and forest; expected also in riparian scrub, mesquite bosque.
LOCALITIES   Lower Sabino Canyon
IDENTIFICATION   A cream-colored central stripe (yellowish toward head), a pale stripe on either side, and dark spots between the stripes in a checkered pattern. Narrow whitish crescent curving upward and forward behind corner of mouth. (Compare with black-necked garter snake.)
Map 24, fig. 42

The checkered garter snake is active at night and less frequently during the day. It is usually seen spring to fall.

Primarily a lowland species and not an inhabitant of mountain canyons,

Figure 42 Checkered garter snake

this snake is much more common along Sabino Creek south of the canyon mouth, where the stream flows across the bajada toward the valley floor. Only occasionally do individuals find their way into the Recreation Area, and our farthest upstream record is a juvenile on the shore of Sabino Lake. The checkered garter snake and the canyon-dwelling black-necked garter snake are found together along the stream in Lower Sabino Canyon.

Checkered and black-necked garter snakes only rarely bite when handled, but instead defend themselves by releasing a foul-smelling fluid from anal glands, as do other garter snakes. Other snakes regularly deploy defensive musk, but garter snake musk is most notably repulsive, and its odor may cling to one's hands for several hours. After being handled and released, a checkered garter snake south of the Recreation Area boundary coiled, flattened its head, and elevated its tail, winding it into a tight spiral in a defensive display.

Groundsnake
Sonora semiannulata

ABUNDANCE	Hypothetical
HABITAT AND LOCALITIES	Expected in semidesert grassland in Upper Sabino Canyon.
IDENTIFICATION	Small, head barely wider than body, variably marked but usually with a dark blotch on each scale. One seen near the Recreation Area had light-colored sides and a broad reddish central stripe interrupted by black bands, but some individuals may lack bands. (Compare with banded sand snake and long-nosed snake.)

Groundsnakes are active both by day and at night. In April 2002 we found an adult groundsnake (see Identification) in Sabino Canyon on the trail into Sabino Basin, 1.2 mi. north of the Recreation Area boundary. It was in semidesert grassland on the eastern canyon slope, elevation approximately 3800 ft., at local sunset, 5:20 p.m. We also have a sight report of a similarly marked groundsnake along the same stretch of trail twenty years earlier, in March 1982, at about 3:00 p.m. We expect this secretive species eventually to be seen along the Phoneline Trail in Upper Sabino Canyon (fig. 16).

The groundsnake has grooved rear teeth and may be mildly venomous, though this has not been well studied. It is not dangerous to people.

Banded Sand Snake
Chilomeniscus cinctus

ABUNDANCE	Uncommonly seen
HABITAT	Bajada desertscrub, paloverde-saguaro desertscrub, riparian scrub; expected also especially in mesquite bosque, and in semidesert grassland.
LOCALITIES	Bajada, foothills, Rattlesnake Canyon, Upper Sabino Canyon; expected in Lower Sabino Canyon, Bear Canyon.
IDENTIFICATION	Small, yellowish, washed with orange down the back, and crossed by jet-black bands. Head no broader than body, snout flattened, lower jaw inset. (Compare with groundsnake.) Map 25, pl. 13

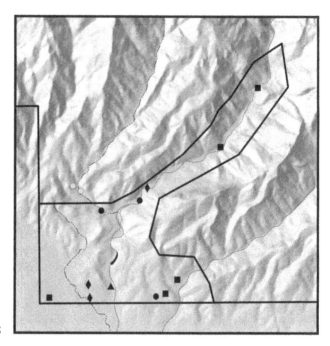

Map 25

- ● Banded sand snake
- ▲ Southwestern black-headed snake
- ■ Western lyresnake
- ◆ Night snake

This nocturnal snake is usually seen March to October. Surface activity may rarely extend into early November.

Because they are adapted to move beneath the surface of loose soil, banded sand snakes live primarily in sandy washes and in loose soil and debris under desert trees, where they feed on invertebrates, especially sand-burrowing cockroaches. In the Tucson region, however, they also inhabit rocky slopes and tend to be quite common in canyons. Thus, the infrequency of sightings likely reflects difficulty in observing a rather common, secretive species. In the Recreation Area and its vicinity these small snakes are seen occasionally on the roads at night, especially in April and May, and we have several records of sand snakes trapped in small plunge pools in Rattlesnake Canyon.

Like the groundsnake, the banded sand snake has grooved rear teeth and may be mildly venomous, though it is not dangerous to people.

Southwestern Black-headed Snake
Tantilla hobartsmithi

ABUNDANCE	Rarely seen
HABITAT	Mesquite bosque; expected also in riparian scrub, riparian woodland and forest, paloverde-saguaro desertscrub, bajada desertscrub.
LOCALITIES	Lower Sabino Canyon; expected in Upper Sabino Canyon, Bear Canyon, Rattlesnake Canyon, foothills, bajada. On the bajada and in the foothills, expected especially in and near washes.
IDENTIFICATION	Small and very slender, with a blackish, flattened head; a pale, narrow neck band; and a brownish, unpatterned back. Central reddish stripe on belly. (Compare with ring-necked snake.) Map 25

The apparent rarity of the southwestern black-headed snake in the Recreation Area is due in part to its secretiveness. This small predator of centipedes, millipedes, and insects spends most of its time underground and beneath rocks and other objects, though it is sometimes seen active on the surface at night. We found this snake only once, in a pitfall trap on the floor of Lower Sabino Canyon.

Like the western lyresnake, this is a mildly venomous snake with enlarged, grooved rear teeth. It is harmless to humans.

Western Lyresnake
Trimorphodon biscutatus

ABUNDANCE	Uncommonly seen
HABITAT	Paloverde-saguaro desertscrub, occasionally bajada desertscrub; expected also in semidesert grassland, mesquite bosque, riparian scrub, riparian woodland and forest.
LOCALITIES	Upper and Lower Sabino Canyon, Bear Canyon, foothills, occasionally on the bajada; expected in Rattlesnake Canyon.
IDENTIFICATION	Head broad and triangular with a forward-pointing V-shaped mark; back with dark, roughly hexagonal

blotches on a light brownish or grayish background, each blotch with a pale border and a pale central bar. Pupil vertical.

Map 25, pl. 14

Primarily inhabitants of rocky environments, where they spend the days hidden in crevices, in the Recreation Area western lyresnakes are usually encountered at night on roads in the canyons and foothills. They are usually seen spring to fall, most often April–June; however, a juvenile once turned up on a sidewalk at the Visitor Center on a rainy February morning.

When disturbed, a lyresnake may coil, vibrate its tail, and hiss. This aggressive behavior, in a snake with a broad head and hexagonal markings, makes it appear strikingly like a black-tailed rattlesnake. The western lyresnake is in fact mildly venomous. It chews its venom into its prey, using a pair of grooved teeth at the rear of its mouth. The venom is not known to be dangerous to humans, though marked local swelling has been reported by valiant herpetologists who have used themselves as experimental subjects.

Night Snake
Hypsiglena torquata

ABUNDANCE	Uncommonly seen
HABITAT	Paloverde-saguaro desertscrub, bajada desertscrub; expected also in riparian woodland and forest, riparian scrub, mesquite bosque, semidesert grassland.
LOCALITIES	Foothills, bajada, Upper Sabino Canyon; expected in Lower Sabino Canyon, Bear Canyon, Rattlesnake Canyon. Known or expected throughout the Recreation Area.
IDENTIFICATION	Small, light brown, with a flattened head. Dark stripe extending through the eye above a whitish upper lip; large, irregularly shaped dark spot or spots on the neck; smaller brown blotches on the back. Pupil vertical. Map 25, fig. 43

Surface activity is nocturnal, but the night snake is also active by day, using a lie-in-wait strategy to catch lizards beneath rocks, logs, and other objects. It is usually seen spring to fall, but also occasionally in winter.

Figure 43 Night snake

This small, secretive snake is most often encountered on the roads after dark. The night snake is a ground-dwelling species that, when threatened, broadens its head at the corners of its mouth, hisses, and makes false, closed-mouth strikes. It is often mistaken for a juvenile rattlesnake (though even a very young rattlesnake has a distinctive button on its tail). Like several other snakes in the Recreation Area, the night snake is a mildly venomous rear-fanged snake, but apparently it is completely harmless to humans.

Ring-necked Snake
Diadophis punctatus

ABUNDANCE	Uncommon
HABITAT	Riparian woodland and forest, mesquite bosque, riparian scrub; expected occasionally near these habitats in semidesert grassland, paloverde-saguaro desertscrub.
LOCALITIES	Upper and Lower Sabino Canyon, Rattlesnake Canyon; expected in Bear Canyon.

IDENTIFICATION   Slender, with an unpatterned bluish gray to olive back
and a bright yellowish or orange neck band. Belly yellow
or orange, grading to red on the tail. (Compare with
southwestern black-headed snake.)
Map 26

In and near the Recreation Area, the ring-necked snake has been seen moving on the surface from March through October. It is active both by day and at night.

In the Santa Catalina Mountains, as elsewhere in southern Arizona, this snake is primarily an inhabitant of higher-elevation grasslands and evergreen woodlands, but it frequents lower elevations in the deciduous riparian woodlands of major drainageways. Its distribution along Sabino Creek today extends well downstream from the canyon mouth (we found it 0.9 mi. south of the Recreation Boundary in 2002), and it was reliably reported near the Santa Cruz River at San Xavier as late as 1940.

A greater roadrunner in Lower Sabino Canyon dropped a ring-necked snake it had caught, then remained nearby for several minutes opening and closing its beak in apparent distaste. The snake had produced a nauseating odor from anal scent glands, and apart from a few beak marks it seemed unharmed. The bright red undersurface of a ring-necked snake's tail may serve to warn potential predators of the animal's unappetizing qualities or perhaps of a hazard at the opposite end—this is a rear-fanged snake with a venom toxic to other reptiles, though probably not dangerous to people.

Like the Sonoran coralsnake, which tends to replace the ring-necked snake in desert environments (although there is considerable overlap at intermediate elevations), this is a snake-eating snake. Prey scarcity probably explains why neither species is abundant.

Sonoran Coralsnake
Micruroides euryxanthus

ABUNDANCE   Rarely seen
HABITAT   Reported mostly in and near riparian communities
AND LOCALITIES   in the mountain canyons, but expected in all biotic
communities and localities in the Recreation Area.
IDENTIFICATION   Slender, glossy, and vividly colored, with broad,
straight-edged red, black, and yellowish bands
encircling the body. Red bands bordered on each side

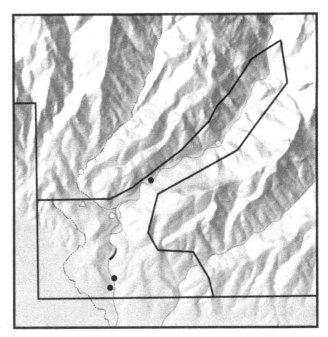

Map 26

●   **Ring-necked snake**

▲   **Sonoran coralsnake**

by yellow; front of head black. (Compare with Sonoran mountain kingsnake.)
Map 26

The Sonoran coralsnake is usually seen during the summer in and near the Recreation Area. Its activity is nocturnal, crepuscular, and occasionally diurnal.

Besides the rattlesnakes, this is the only snake in the Recreation Area known to be potentially dangerous to people. However, because of its rarity, small size, and small teeth, the Sonoran coralsnake poses very little threat to visitors, despite its highly potent venom. We did not encounter coralsnakes in our fieldwork, but we have received sight reports from Upper Sabino Canyon, Rattlesnake Canyon, and Bear Canyon. There are also several vouchered museum records (though with a less precisely defined "Sabino Canyon" locality), as would be expected, given the Recreation Area's productive desert and semidesert canyon environments. (See also ring-necked snake.)

Western Diamondback Rattlesnake
Crotalus atrox

ABUNDANCE   Common
HABITAT   Bajada desertscrub, paloverde-saguaro desertscrub,
mesquite bosque; expected also in riparian scrub,
riparian woodland and forest.
LOCALITIES   Bajada, foothills, Lower Sabino Canyon, Bear Canyon
in the vicinity of Lower Bear Picnic Area. Also a single
report in Upper Sabino Canyon (see below).
IDENTIFICATION   Large, tan or grayish overall, with roughly diamond-
shaped blotches partly edged with white, and many
dark speckles. Tail with conspicuous alternating black
and white rings of similar width. (Compare with
Mohave rattlesnake.)
Map 27, fig. 44, pl. 15

Western diamondbacks are usually seen moving March to October, but they are expected to be active around dens during other months, especially in late winter. They are mostly nocturnal, but are also markedly diurnal during fall, winter, and spring. Mating probably occurs mainly in spring and fall; it has been observed in late March and early September. Newborns, with a single button on their tails, first appear in August; the rapidly growing juveniles may greatly outnumber adults on the roads for the remainder of the season.

Within the Recreation Area the western diamondback is primarily a resident of the bajada, but it also is found in the gently sloping foothills bordering the bajada and on the broad floors of the lower reaches of Sabino Canyon and Bear Canyon. We have not seen this snake in Upper Sabino Canyon, though we have an unconfirmed report near the Rattlesnake Canyon confluence—a plausible record by analogy with the local distribution of the desert spiny lizard (see that account). If the western diamondback inhabits narrow canyon environments in the Recreation Area, its numbers are greatly exceeded there by the black-tailed rattlesnake and the tiger rattlesnake.

Like the other rattlesnakes, this species poses only a limited threat to visitors who do not molest it. When encountered in the open during the day (the circumstances in which it is most often noticed), a western diamondback does not always rattle or hiss, but almost invariably retreats immediately to the protection of a rock or shrub, where it remains coiled

Figure 44 Juvenile western diamondback rattlesnake

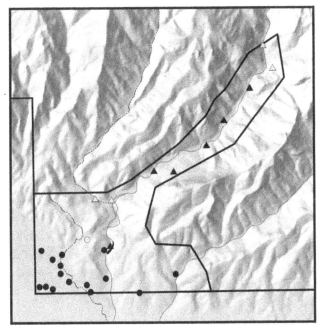

Map 27

●   **Western diamondback rattlesnake**

▲   **Black-tailed rattlesnake**

until the observer departs. Nevertheless, visitors should be aware of the presence of rattlesnakes in all environments in the Recreation Area and should exercise caution at any time of day or night (see Safety for Herptiles and Humans).

Black-tailed Rattlesnake
Crotalus molossus

ABUNDANCE	Common
HABITAT	Riparian woodland and forest, riparian scrub, mesquite bosque, paloverde-saguaro desertscrub, desertscrub/grassland ecotone; expected in semidesert grassland.
LOCALITIES	Upper and Lower Sabino Canyon, Bear Canyon, Rattlesnake Canyon; expected occasionally in foothills.
IDENTIFICATION	Yellowish or greenish, marked with brown cross-bands or hexagons, these with lighter edges and centers. Tail dark gray or blackish, often with still darker rings faintly visible; snout often blackened. Map 27, pl. 16

Black-tailed rattlesnakes are active both day and night. They are usually seen May to September, but they may sometimes be active in fall and winter as well. Visitors probably encounter them most frequently on summer mornings, crossing the road in Upper Sabino Canyon (fig. 7). Young-of-the-year are seen very infrequently, unlike young western diamondbacks.

Usually less aggressive when disturbed than a western diamondback rattlesnake, a black-tailed rattlesnake may coil and rattle but less often strikes. Because of this snake's overall golden or slightly greenish color, visitors sometimes misidentify it as a "green Mohave rattlesnake."

The black-tail is a mountain form that is primarily a canyon-dweller in the Recreation Area. It is apparently less abundant in the rocky foothills than the tiger rattlesnake, which shares many of its narrow canyon environments. The distribution of the black-tailed rattlesnake overlaps that of the western diamondback in the broad lower reaches of Sabino Canyon and Bear Canyon.

Mohave Rattlesnake
Crotalus scutulatus

ABUNDANCE   Rare or possibly no longer present
HABITAT   Possibly a rare inhabitant of bajada desertscrub.
IDENTIFICATION   Usually greenish to brownish, with a row of distinct, roughly diamond-shaped dark blotches outlined by lighter scales. Tail with conspicuous alternating light and dark rings, the dark rings narrower than the light. (Compare with western diamondback and black-tailed rattlesnakes.)

The Mohave rattlesnake is nocturnal when days are hot but diurnal during cooler seasons. It is primarily a resident of flatlands and bottomlands; the uppermost bajada, as in the Recreation Area, is on the fringe of its habitat. There is a 1942 specimen from "Sabino Cañon" (American Museum of Natural History AMNH 64349), which was probably taken in the Recreation Area, and a 1963 specimen approximately 1 mi. south of the Sabino Canyon Visitor Center (University of Arizona UAZ 27783). It is likely that individual Mohave rattlesnakes occasionally reach the Recreation Area or did so before urbanization decimated the local population.

Any rattlesnake found in the Recreation Area having a tail marked with alternate light and dark bands should be carefully observed (from a safe distance) in case it proves to be this species rather than the western diamondback. "Green Mohave rattlesnakes" sometimes reported in Upper Sabino Canyon are actually black-tailed rattlesnakes.

Tiger Rattlesnake
Crotalus tigris

ABUNDANCE   Common
HABITAT   Paloverde-saguaro desertscrub, semidesert grassland, mesquite bosque, riparian scrub; occasionally in bajada desertscrub; expected also in riparian woodland and forest.
LOCALITIES   Upper Sabino Canyon, Lower Sabino Canyon, Rattlesnake Canyon, foothills; expected in Bear Canyon. Occasionally extreme margins of bajada bordering foothills.

IDENTIFICATION   Grayish to pinkish or somewhat orange, with poorly defined dark cross-bands rather than blotches, and with similar rings on the tail. Head small; tail often brownish near the large rattle.
Map 28, fig. 45

Tiger rattlesnakes are active both by day and at night. They are usually seen May to November, most frequently in August and September. We found a newborn in early August.

Usually not overly aggressive when encountered, tiger rattlesnakes often fail to rattle and seldom strike. (One found active during the day on a rocky slope in Upper Sabino Canyon retreated under a bush, coiled, and tucked in both its head and its tail.) However, bite they will, and their venom is highly toxic.

In the Recreation Area the tiger rattlesnake is apparently more broadly distributed than the black-tailed rattlesnake. Although both species inhabit the mountain canyons, the tiger rattlesnake is seen more frequently than the black-tail in the rocky foothills. The distribution of the tiger rattlesnake overlaps that of the western diamondback in the foothills, in Lower Sabino Canyon, and (we expect) in the vicinity of the former Lower Bear Picnic Area.

Figure 45 Tiger rattlesnake in defensive posture, head and rattle inside coil

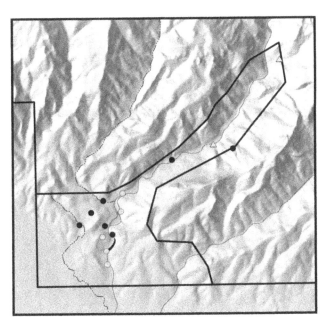

Map 28

● Tiger rattlesnake

▲ Arizona black rattlesnake

Arizona Black Rattlesnake
Crotalus cerberus

ABUNDANCE	Rarely seen
HABITAT	Semidesert grassland; reported near riparian communities on the floor of Upper Sabino Canyon.
LOCALITIES	Upper Sabino Canyon
IDENTIFICATION	Very dark gray or brown with a row of large, still darker blotches separated by narrow light cross-bands, the blotches grading to rings on the tail. Some snakes black by day, with only the light cross-bands visible, but lightening and showing the blotches at night. Map 28

Activity is both diurnal and nocturnal. Juveniles are more lightly colored than adults and develop the characteristic dark coloration of the species as they mature.

The Arizona black rattlesnake is one of several snakes and lizards that in the mountains of southeastern Arizona live primarily in higher-elevation woodlands and forests, but whose distributions extend downslope into desert canyon communities. We have not encountered this rattlesnake in the Recreation Area, but we have several reliable sight reports on and near the road in Upper Sabino Canyon. A specimen or photograph would be valuable to document its presence in the Recreation Area.

Appendix A Scientific Names of Plants and Animals

Plants

African sumac	*Rhus lancea*
Arizona rosewood	*Vauquelinia californica*
Arizona sycamore	*Platanus wrightii*
Arizona walnut	*Juglans major*
Barrel cactus	*Ferocactus wislizeni*
Bermuda grass	*Cynodon dactylon*
Blue paloverde	*Parkinsonia florida*
Bonpland willow	*Salix bonplandiana*
Brittlebush	*Encelia farinosa*
Buffelgrass	*Cenchrus ciliaris*
Bermuda grass	*Cynodon dactylon*
Burrobrush	*Hymenoclea salsola*
Burroweed	*Isocoma tenuisecta*
Canyon grape	*Vitis arizonica*
Canyon ragweed	*Ambrosia ambrosioides*
Catclaw	*Acacia greggii*
Common cocklebur	*Xanthium strumarium*
Coursetia	*Coursetia glandulosa*
Creosotebush	*Larrea divaricata*
Deer grass	*Muhlenbergia rigens*
Desert broom	*Baccharis sarothroides*
Desert hackberry	*Celtis pallida*
Desert mistletoe	*Phoradendron californicum*
Emory oak	*Quercus emoryi*
Engelmann prickly pear	*Opuntia engelmannii*
Fairy duster	*Calliandra eriophylla*
Foothill paloverde	*Parkinsonia microphylla*
Fountain grass	*Pennisetum setaceum*
Fremont cottonwood	*Populus fremontii*
Giant reed	*Arundo donax*
Goodding willow	*Salix gooddingii*
Gum bumelia	*Sideroxylon lanuginosum*
Hop bush	*Dodonaea viscosa*
Jumping cholla	*Opuntia fulgida*
Lehmann lovegrass	*Eragrostis lehmanniana*

London rocket	*Sisymbrium irio*
Mexican blue oak	*Quercus oblongifolia*
Natal grass	*Melinis repens*
Net leaf hackberry	*Celtis reticulata*
Ocotillo	*Fouquieria splendens*
Rabbitfoot grass	*Polypogon monspeliensis*
Red brome	*Bromus rubens*
Sacred datura	*Datura wrightii*
Saguaro	*Carnegiea gigantea*
Sotol	*Dasylirion wheeleri*
Southern cattail	*Typha domingensis*
Staghorn cholla	*Opuntia versicolor*
Sweet resinbush	*Euryops multifidus*
Tamarisk	*Tamarix ramosissima*
Teddy bear cholla	*Opuntia bigelovii*
Triangle bur-sage	*Ambrosia deltoidea*
Turpentine bush	*Ericameria laricifolia*
Velvet ash	*Fraxinus velutina*
Velvet mesquite	*Prosopis velutina*
Wait-a-minute bush	*Mimosa aculeaticarpa* var. *biuncifera*
White thorn	*Acacia constricta*
Wright lippia	*Aloysia wrightii*

Animals Other Than Amphibians and Reptiles

Bluegill	*Lepomis macrochirus*
Cactus wren	*Campylorhynchus brunneicapillus*
Cassin's kingbird	*Tyrannus vociferans*
Gila chub	*Gila intermedia*
Gila topminnow	*Poeciliopsis occidentalis occidentalis*
Gray fox	*Urocyon cinereoargenteus*
Greater roadrunner	*Geococcyx californianus*
Green sunfish	*Lepomis cyanellus*
Hooded oriole	*Icterus cucullatus*
Largemouth bass	*Micropterus salmoides*
Longfin dace	*Agosia chrysogaster*
Mosquitofish	*Gambusia affinis*
Northern crayfish	*Orconectes virilis*
Rock squirrel	*Spermophilus variegatus*
Round-tailed ground squirrel	*Spermophilus tereticaudus*
White-throated woodrat	*Neotoma albigula*

Appendix B Photographic Stations

Geographic coordinates of the photographic stations were determined using a hand-held global positioning system unit. Datum is WGS84/NAD83. Most positions are accurate within 26 ft. or less. All photos were taken on 35-mm film with fixed focal-length lenses.

Figure	Latitude (°N)	Longitude (°W)	Focal length (mm)
2a, 2b	32.31712	110.81785	50
3a, 3b	32.31978	110.81690	50
4a, 4b	32.32532	110.80703	50
5	32.31782	110.80950	28
6	32.31754	110.80940	50
7	32.33321	110.79076	50
8	32.32231	110.81206	28
9	32.31046	110.80015	50
10	32.31347	110.79716	50
11a, 11b	32.31155	110.82005	50
12	32.30951	110.82131	28
13	32.31521	110.81857	50
14	32.33427	110.78977	50
15	32.33427	110.78977	50
16[a]	32.33776	110.78055	50
17a, 17b	32.33636	110.78765	28
18a, 18b[b]	32.31531	110.81131	28
19	32.31071	110.81068	50
20	32.31268	110.79755	28
21	32.31429	110.81144	28
22a, 22b	32.31072	110.81110	50
23	32.31694	110.81084	50

[a]Position is accurate within 66 ft.
[b]Position is for figure 18b. See figure legend.

Appendix C English to Metric Unit Conversion

The formulae below are to aid in converting measurements expressed in English units to the International System of Units (metric system).

Linear

        1 inch (in.) = 2.54 centimeters
        1 foot (ft.) = 0.305 meter
        1 yard (yd.) = 0.914 meter
        1 mile (mi.) = 1.61 kilometer

Area

        1 acre = 0.405 hectare
        1 square mile = 2.59 square kilometers

Flow rates

        1 cubic foot per second (cfs) = 0.0283 cubic meters per second

Temperature

        $°C = (5/9)(°F - 32)$

Appendix D Sabino Creek Timeline

This simplified chronology summarizes important events in the environmental history of Sabino Creek since the beginning of the twentieth century. Unless otherwise stated, entries refer to the portion of the stream within the Sabino Canyon Recreation Area. Earliest and latest specimens are based on collection records available to the authors and may not represent extreme dates of occurrence. We have omitted certain exotic fishes that have occurred primarily below the National Forest boundary and have listed only floods exceeding 4000 cubic feet per second (cfs).

1906	Lowland leopard frog reported abundant; earliest collection specimens. (Will remain common or abundant at least into mid-1960s.)
1917	Native Gila chub, Gila topminnow, and longfin dace all collected by this date, though some specimens possibly below the National Forest boundary. (Historical occurrence of topminnow and dace in Recreation Area likely but uncertain.)
1934–1937	Construction of stone bridges in Upper and Lower Sabino Canyon, as well as dam at Sabino Lake.
1939	Sabino Lake opened for fishing. (Stocked at least into mid-1950s with bass, bluegill, sometimes trout, but of these only largemouth bass will eventually become established, mostly below the National Forest boundary.)
1941	Sabino Lake drained, sediment and vegetation removed.
1943	Earliest mosquitofish specimens recorded (possibly below the National Forest boundary—presence in the Recreation Area by this date uncertain).
1947	Latest Gila topminnow specimens recorded (possibly below the National Forest boundary).
1949	Latest longfin dace specimens recorded (possibly below the National Forest boundary).
1950	Sabino Lake perhaps dredged (evidence uncertain).
1954 Mar	Flood, 5110 cfs.

1955	Northern crayfish present below the National Forest boundary by this date (presence in Recreation Area uncertain).
1959 Jul	Flood, 4240 cfs.
1960	Earliest American bullfrog specimens recorded. (Crayfish probably in Recreation Area by this date.)
1964	Green sunfish present below National Forest boundary by this date (presence in Recreation Area uncertain).
1965–1966	American bullfrog common in Sabino Lake, specimens outnumbering lowland leopard frog in collections.
1966 Aug	Flood, 6400 cfs. Local bullfrog reproduction ceases (tadpoles absent from record until 1977).
1967	Latest bullfrog collection specimen (but a reduced population will wax and wane up to the present).
1968	Latest lowland leopard frog specimens recorded (Lower Sabino Canyon).
1970 Sep	Flood, 7730 cfs.
1972	Leopard frogs reported in and near Lower Sabino Canyon (last record in Recreation Area until 2000).
1977–1978	Bullfrog tadpoles observed in Lower Sabino Canyon.
1978 Dec	Flood, 7400 cfs. Bullfrog reproduction ceases (tadpoles absent until 1989).
1979	Earliest definite green sunfish record in Recreation Area (near Rattlesnake Creek confluence), but species almost certainly present earlier in decade.
1981	Surveys record leopard frogs only above Recreation Area.
1982	Fish distributions mapped. Gila chub above Rattlesnake Creek confluence, mosquitofish below confluence. Green sunfish up to second bridge, Upper Sabino Canyon (its distribution will expand upstream until 1995).
1983 Oct	Flood, 6500 cfs.
1989–1992	Bullfrog tadpoles and juveniles observed in Sabino Lake.
1991–1992	Several leopard frog sightings in Upper Sabino Canyon.
1993 Jan	Flood, 12,900 cfs. Mosquitofish extirpated. Bullfrog reproduction ceases (tadpoles not recorded again as of 2005).

1995 Jun	Sabino Fire on southeastern slope of Upper Sabino Canyon. Minor sedimentation and algal bloom during ensuing monsoon.
1995 Oct	Green sunfish recorded as far upstream as highest bridge, Upper Sabino Canyon. Gila chub numbers greatly reduced within distribution of sunfish.
1999 Jun	Stream from highest bridge in Upper Sabino Canyon down to Sabino Lake treated with chemicals to eliminate sunfish.
1999 Jul	Flood, 15,400 cfs, the largest recorded.
2000	Abundant lowland leopard frog juveniles observed in Sabino Lake.
2001–2003	No leopard frogs found in surveys to 3840 ft.
2003 Jun	Aspen Fire burns much of Sabino Creek watershed. Approximately 1000 Gila chub removed from Upper Sabino Canyon as a precaution against anticipated flooding.
2003 Jul	Flash flooding begins with onset of monsoon rains, decimating remaining Gila chub as well as northern crayfish, extirpating remaining green sunfish (below Sabino Lake), and filling many formerly perennial pools with sediment.
2005 May	Approximately 350 of the Gila chub removed in 2003 reintroduced into Upper Sabino Canyon.

Appendix E Checklist of Amphibians and Reptiles in the Sabino Canyon Recreation Area

E   extirpated
H   hypothetical
I   introduced

Amphibians

Salamanders

Tiger salamander (I)   *Ambystoma tigrinum*

Toads and Frogs

Couch's spadefoot   *Scaphiopus couchii*
Mexican spadefoot (H)   *Spea multiplicata*
Sonoran Desert toad   *Bufo alvarius*
Red-spotted toad   *Bufo punctatus*
Great Plains toad   *Bufo cognatus*
Canyon treefrog   *Hyla arenicolor*
Lowland leopard frog (E?)   *Rana yavapaiensis*
American bullfrog (I)   *Rana catesbeiana*

Reptiles

Turtles

Sonoran mud turtle   *Kinosternon sonoriense*
Desert tortoise   *Gopherus agassizii*
Western box turtle (E?, I?)   *Terrapene ornata*
Eastern box turtle (I)   *Terrapene carolina*
Pond slider (I)   *Trachemys scripta*
Painted turtle (I)   *Chrysemys picta*

Lizards

Western banded gecko   *Coleonyx variegatus*
Mediterranean gecko (H, I)   *Hemidactylus turcicus*
Eastern collared lizard   *Crotaphytus collaris*
Long-nosed leopard lizard   *Gambelia wislizenii*
Zebra-tailed lizard   *Callisaurus draconoides*
Greater earless lizard   *Cophosaurus texanus*
Clark's spiny lizard   *Sceloporus clarkii*

Desert spiny lizard	*Sceloporus magister*
Side-blotched lizard	*Uta stansburiana*
Tree lizard	*Urosaurus ornatus*
Regal horned lizard	*Phrynosoma solare*
Great Plains skink	*Eumeces obsoletus*
Canyon spotted whiptail	*Aspidoscelis burti*
Sonoran spotted whiptail	*Aspidoscelis sonorae*
Tiger whiptail	*Aspidoscelis tigris*
Madrean alligator lizard (H)	*Elgaria kingii*
Gila monster	*Heloderma suspectum*

Snakes

Western blind snake	*Leptotyphlops humilis*
Sonoran whipsnake	*Masticophis bilineatus*
Coachwhip	*Masticophis flagellum*
Saddled leaf-nosed snake	*Phyllorhynchus browni*
Western patch-nosed snake	*Salvadora hexalepis*
Graham patch-nosed snake (H)	*Salvadora grahamiae*
Glossy snake	*Arizona elegans*
Gopher snake	*Pituophis catenifer*
Common kingsnake	*Lampropeltis getula*
Sonoran mountain kingsnake (H)	*Lampropeltis pyromelana*
Long-nosed snake	*Rhinocheilus lecontei*
Black-necked garter snake	*Thamnophis cyrtopsis*
Checkered garter snake	*Thamnophis marcianus*
Groundsnake (H)	*Sonora semiannulata*
Banded sand snake	*Chilomeniscus cinctus*
Southwestern black-headed snake	*Tantilla hobartsmithi*
Western lyresnake	*Trimorphodon biscutatus*
Night snake	*Hypsiglena torquata*
Ring-necked snake	*Diadophis punctatus*
Sonoran coralsnake	*Micruroides euryxanthus*
Western diamondback rattlesnake	*Crotalus atrox*
Black-tailed rattlesnake	*Crotalus molossus*
Mohave rattlesnake	*Crotalus scutulatus*
Tiger rattlesnake	*Crotalus tigris*
Arizona black rattlesnake	*Crotalus cerberus*

References

Behler, J. L., and F. W. King. 1979. *The Audubon Society Field Guide to North American Reptiles and Amphibians*. New York: Chanticleer Press.

Brown, D. E., C. H. Lowe, and C. P. Pase. 1980. *A Digitized Systematic Classification for Ecosystems with an Illustrated Summary of the Natural Vegetation of North America*. U.S. Department of Agriculture Forest Service General Technical Report RM-73.

Hare, T. 1995. *Poisonous Dwellers of the Desert*. Tucson: Western National Parks Association.

King, F. W. 1932. Herpetological records and notes from the vicinity of Tucson, Arizona, July and August, 1930. *Copeia* 1932:175-7.

Lazaroff, D. W. 1993. *Sabino Canyon: The Life of a Southwestern Oasis*. Tucson: University of Arizona Press.

———. 1998. *Arizona-Sonora Desert Museum Book of Answers*. Tucson: Arizona-Sonora Desert Museum Press.

Lowe, C. H. 1964. Amphibians and reptiles of Arizona. In *The Vertebrates of Arizona*, ed. C. H. Lowe, 153-74. Tucson: University of Arizona Press.

Lowe, C. H., and J. W. Wright. 1964. Species of the *Cnemidophorus exsanguis* subgroup of whiptail lizards. *Journal of the Arizona Academy of Science* 3(2):78-80.

Lowe. C. H., C. R. Schwalbe, and T. B. Johnson. 1986. *The Venomous Reptiles of Arizona*. Phoenix: Arizona Game and Fish Department.

McCoy, C. V. 1932. Herpetological notes from Tucson, Arizona. *Occasional Papers of the Boston Society of Natural History* 8:11-24.

National Oceanic and Atmospheric Administration. 1982. *Monthly Normals of Temperature, Precipitation, and Heating and Cooling Degree Days 1951-80 Arizona*. Asheville, N.C.: Environmental Data and Information Service, National Climatic Center.

Rosen, P. C., and C. R. Schwalbe. 2002. Widespread effects of introduced species on reptiles and amphibians in the Sonoran Desert region. In *Invasive Exotic Species in the Sonoran Region*, ed. B. Tellman, 220-40. Tucson: University of Arizona Press.

Ruthven, A. G. 1907. A collection of reptiles and amphibians from southern New Mexico and Arizona. *Bulletin of the American Museum of Natural History* 23:483-603.

Smith, H. M., and E. D. Brodie Jr. 1982. *Reptiles of North America*. New York: St. Martin's Press.

Stebbins, R. C. 2003. *A Field Guide to Western Reptiles and Amphibians*. 3rd ed. Boston: Houghton Mifflin Co.

Turner, R. M., R. H. Webb, J. E. Bowers, and J. R. Hastings. 2003. *The Changing Mile Revisited: An Ecological Study of Vegetation Change with Time in the Lower Mile of an Arid and Semiarid Region.* Tucson: University of Arizona Press.

Van'Denburgh, J., and J. R. Slevin. 1913. A list of the amphibians and reptiles of Arizona, with notes on the species in the collection of the Academy. *Proceedings of the California Academy of Sciences,* 4th series, 3:391–454.

Whittaker, R. H., and W. A. Niering. 1965. Vegetation of the Santa Catalina Mountains, Arizona: A gradient analysis of the south slope. *Ecology* 46(4):429–52.

Index

About the Authors

David W. Lazaroff began studying Sabino Canyon while working as an environmental education specialist for Coronado National Forest from 1977 to 1986. His book, *Sabino Canyon: The Life of a Southwestern Oasis*, was published in 1993. He is now an independent naturalist, writer, and photographer.

Philip C. Rosen has spent twenty-one years studying the ecology and conservation biology of Arizona's aquatic, riparian, and desert amphibians and reptiles, as a graduate student, first at Arizona State University and then at the University of Arizona, where he is now a research scientist in the School of Natural Resources.

Charles H. Lowe Jr. (1920–2002) arrived in 1950 at the University of Arizona, where he led a colorful and storied existence as a leading Southwestern natural historian, professor, and herpetologist. He is widely known for his book *The Vertebrates of Arizona* and his contributions to studies of the biotic communities of the Southwest and the ecology of the saguaro. He described many new species and subspecies of amphibians and reptiles and, with his students, studied ecological physiology and was central to solving the riddle of the all-female whiptail lizards.

The Southwest Center Series